Next Generation Artificial Intelligence Driven Smart and Renewable Energy

To provide for a sustainable future, the potential synergies at the dynamic intersection of renewable energy (RE) incorporated with smart energy and artificial intelligence (AI) must be exploited. RE is crucial to preserve the environment. Energy involving various systems must be optimized and assessed to provide better performance. However, the design and development of RE systems remain a challenge. Advanced optimization techniques, AI, and machine learning (ML) play a crucial role in implementing the latest innovative research in the field of renewable energy-integrated electrical systems. This book also describes the practical challenges encountered and the solutions and future scope to be adopted. Applications of a variety of advanced optimization and AI techniques in the design and development of RE-integrated systems are discussed to provide new solutions in the RE domain.

Key features:

- Discusses modern modeling/control approaches for improving renewable energy integrating artificial intelligence-driven power systems.
- Describes the principles and methods of renewable energy generation technologies and an analysis of their implementation, management, optimization, and related economic advantages.
- Presents critical information on the technological design and policy issues that must be considered while implementing a smart grid.
- Explains the metaheuristic optimization algorithm for complex electrical systems and the whale optimization algorithm-based multi-objective hydrothermal scheduling.
- Covers the electric vehicle charging station in the distribution network and transient stability constraint optimal power flow problem using chaotic quasi-oppositional chemical reaction optimization.

The topics covered include microgrids, wind power, solar photo voltaic (PV), optimal power flow (OPF), grid-connected inverter, electric vehicle, combined heat and power economic dispatch, FACTS tools for smart energy, harmonic impedance of a salient pole synchronous generator (HI), maximum power point tracking (MPPT) and advanced optimization techniques. *Next Generation Artificial Intelligence Driven Smart and Renewable Energy* is ideal for academicians, practitioners, teachers, engineers, industry professionals, researchers, and students in diverse fields, including electrical engineering, electronics and communications engineering, energy, and environmental engineering.

Intelligent Data-Driven Systems and Artificial Intelligence
Series Editor–Harish Garg

Cognitive Machine Intelligence: Applications, Challenges, and Related Technologies
Inam Ullah Khan, Salma El Hajjami, Mariya Ouaissa, Salwa Belqziz, and Tarandeep Kaur Bhatia

Artificial Intelligence and Internet of Things based Augmented Trends for Data Driven Systems
Anshu Singla, Sarvesh Tanwar, and Pao-Ann Hsiung

Modelling of Virtual Worlds Using the Internet of Things
Edited by Simar Preet Singh and Arun Solanki

Data-Driven Technologies and Artificial Intelligence in Supply Chain Tools and Techniques
Mahesh Chand, Vineet Jain, and Puneeta Ajmera

For more information about this series, please visit: www.routledge.com/Intelligent-Data-Driven-Systems-and-Artificial-Intelligence/book-series/CRCIDDSAAI

Next Generation Artificial Intelligence Driven Smart and Renewable Energy

Edited by

Provas Kumar Roy, Sunanda Hazra, and
Chandan Paul

CRC Press
Taylor & Francis Group
Boca Raton London New York

CRC Press is an imprint of the
Taylor & Francis Group, an **informa** business

Front cover image: Sammy666888/Shutterstock

First edition published 2025
by CRC Press
2385 NW Executive Center Drive, Suite 320, Boca Raton FL 33431

and by CRC Press
4 Park Square, Milton Park, Abingdon, Oxon, OX14 4RN

CRC Press is an imprint of Taylor & Francis Group, LLC

ISBN: 978-1-032-76156-5 (hbk)
ISBN: 978-1-041-04247-1 (pbk)
ISBN: 978-1-003-62827-9 (ebk)

DOI: 10.1201/9781003628279

Typeset in Sabon
by codeMantra

Contents

Preface

Recently, AI-driven renewable energy plays a vital role around the world in electrical energy system to make the system smart.

Chapter 1 explores the emerging trends and innovations at the crossroads of AI and renewable energy, including advanced energy storage technologies, decentralized energy systems, and the integration of Internet of Things (IoT) devices.

Chapter 2 tracks the maximum power point of the photovoltaic module so that very much possible power can be extracted from the photovoltaic/solar cell.

Chapter 3 discusses the concept of harmonic impedance because depending on the sequence of switching of the power electronic inverter, the nature of the non-sinusoidal output of the inverter is decided.

Chapter 4 describes how soft computing techniques are the pillars of intelligent systems when applied to real-life engineering problems like PID controller design, finding the maximum power transfer point, controllers for hybrid power distribution grids, and power generation forecasting and usage.

Chapter 5 discusses the valve point effect. It also discusses the reduction of gearbox losses using wind and solar energy which are further related to FACTS devices TCSC and TCPS.

Chapter 6 presents an analysis of a power system-based combined heat and power economic dispatch (CHPED) problem utilizing the quantum whale optimization algorithm (QWOA).

Chapter 7 introduces the modeling and fabrication of electric vehicle for multicriteria framework-based racing frontier ecosystems.

Chapter 8 illustrates the innovative utilization of artificial intelligence (AI) in wind energy generation, specifically focusing on the strategic installation of wind farms near railway tracks and their autonomous control through AI systems.

Chapter 9 presents issues related to the governance and empowerment of independent power producers (IPPs).

Chapter 10 describes a grid-connected inverter system to connect the electricity production for small-scale applications in a single-phase system.

Further, a suitable current controller must be designed to enhance the system's tracking execution and reduce total harmonic distortion (THD).

Chapter 11 addresses the transient stability constraint optimal power flow problem by utilizing the concept of chaotic quasi-oppositional differential search algorithm (CQODSA).

Chapter 12 illustrates AI-powered renewable energy systems for a sustainable future.

Acknowledgments

We would like to thank the reviewers for taking the time and effort to review various chapters of this book. We sincerely appreciate all valuable comments and suggestions, which helped us to improve the quality of the book. Names of the reviewers are given below.

List of reviewers

1. **Dr. Vidyasagar Bhattacharya,** Department of Mechanical Engineering, NHIT, West Bengal, India
2. **Dr. Dipayan Guha,** Electrical Engineering Department, Motilal Nehru National Institute of Technology Allahabad, India
3. **Mr. Dhiman Banerjee,** Department of Electrical Engineering, Kalyani Government Engineering College, West Bengal, India
4. **Mrs. Sriparna Bhattacharya,** Department of Electrical Engineering, Kalyani Government Engineering College, West Bengal, India
5. **Dr. Falguni Chakraborty,** Department of MCA, Dr. B. C. Roy Engineering College, West Bengal, India
6. **Dr. Tsung-Jung Hsieh,** Center of General Education, National Formosa University, Taiwan
7. **Dr. Sanjoy Kumar Saha,** Department of Electrical Engineering, Dr. B. C. Roy Engineering College, West Bengal, India
8. **Dr. Soumen Biswas,** Department of Electrical Engineering, Dr. B. C. Roy Engineering College, West Bengal, India

Also, we would like to acknowledge the publisher, authors, and support staff at Taylor & Francis in the various stages of developing and producing the book.

About the editors

Provas Kumar Roy received a BE degree in Electrical Engineering from R. E. College, Durgapur, Burdwan, India, in 1997; an ME degree in Electrical Machine from Jadavpur University, Kolkata, India, in 2001; and a PhD from NIT Durgapur in 2011. Presently, he is working as a Professor in the Department of Electrical Engineering, Kalyani Government Engineering College, Kalyani, India. He has published more than 140 research papers in international journals, 65 conference paper, and 15 book chapters, and Scopus citation is nearly 4700 with an H-index of 40. His research interests include economic load dispatch, optimal power flow, FACTS, unit commitment, radial distribution system, state estimation, automatic generation control, power system stabilizer, and evolutionary computing techniques. He has supervised 12 PhD scholars in the domain of power system analysis with high penetration of renewable energy and image processing. He has been placed among the "World Ranking of Top 2% Scientists" for the last three consecutive years in the field of energy by Stanford University scientists. He has been awarded many times with outstanding reviewer awards, top scientist awards, top peer reviewer awards, and best paper awards. He has also contributed as a committee member of several national and international conferences. He is a member of the Institute of Engineers and the Indian Society of Technical Education (ISTE). He is a regular reviewer of leading journals, including Elsevier, IET, and IEEE Transactions on Communications.

Sunanda Hazra received a PhD degree in Electrical Engineering. He is presently associated with the Electrical Engineering Department of Haldia Institute of Technology, Haldia, WB, India. He has ten years of teaching experience. He has published around 20 research papers in international journals and conference records. His research interests include load dispatch, hydrothermal scheduling, renewable energy, and optimization techniques. He was awarded the Young Scientist award from VD Good Technology Factory in 2021. He has reviewed a few research works submitted to national/international journals. He has attended and organized several short-term courses/faculty development programs. He is a member of IEI.

Chandan Paul received a BTech degree in Electrical Engineering from Dr. B. C. Roy Engineering College, Durgapur (under West Bengal University of Technology), India, in 2006 and an MTech degree in Electrical Engineering (specialization of Electrical System) from the National Institute of Technology, Durgapur, West Bengal, India, in 2012. He obtained a PhD in Electrical Engineering from IIT (ISM), Dhanbad, in 2022. He has published eight journals in reputed SCI and Scopus-indexed journals. His area of research includes hydrothermal scheduling, optimal power flow, combined heat and power dispatch, and evolutionary algorithms. He has published in three international journals. He is working as an Assistant Professor in the Department of Electrical Engineering, Dr. B. C. Roy Engineering College, Durgapur, India.

List of contributors

Arpan Adhikary
Department of Computer Science and Engineering (AI & ML)
Haldia Institute of Technology
Haldia, West Bengal, India

Jatin Anand
Electrical Engineering Department
Asansol Engineering College
Asansol, India

Ranjit Barua
Centre for Healthcare Science and Technology
Indian Institute of Engineering Science and Technology
Howrah, West Bengal, India

Sunetra Bhomik
Electrical Engineering Department
Asansol Engineering College
Asansol, India

Adhir Baran Chattopadhyay
Electrical Engineering Department
Budge Budge Institute of Technology
Kolkata, India

Samir Das
Biomaterials and Tissue Engineering Lab
School of Medical Science and Technology Indian Institute of Technology
Kharagpur, India

Sudipto Datta
Department of Materials Engineering
Indian Institute of Science
Bangalore, India

Nirmalya Dey
Electrical Engineering Department
Dr. B. C. Roy Engineering College
Durgapur, India

Prasenjit Dey
Electrical Engineering Department
National Institute of Technology
Rourkela, Odisha, India

Susanta Dutta
Electrical Engineering Department
Dr. B. C. Roy Engineering College
Durgapur, India

Arko Ghosal
Electrical Engineering Department
Dr. B. C. Roy Engineering College
Durgapur, India

Arnab Ghosh
Department of Electrical Engineering
National Institute of Technology
Rourkela, Odisha, India

Dipanwita Ghosh
Department of Computer Science and Engineering
Bengal College of Engineering and Technology
Durgapur, West Bengal, India

Pramit Ghosh
Electrical Engineering Department
Dr. B. C. Roy Engineering College
Durgapur, WB, India

Sunanda Hazra
Department of Electrical Engineering
Haldia Institute of Technology
Haldia, West Bengal, India

Nipun Jain
Department of Materials Engineering
Indian Institute of Science
Bangalore, Karnataka, India

Kaushik Kumar
Department of Mechanical Engineering
Birla Institute of Technology
Mesra, Ranchi, India

Jibendu Kumar Mantri
Department of Computer Science and Applications
MSCB University
Baripada, Odisha, India

Pfano Mashau
Department of Mechanical Engineering
Durban University of Technology Business School
Durban, South Africa

Sudeshna Nath
Electrical Engineering Department
Haldia Institute of Technology
Haldia, West Bengal, India

Asit Kumar Nayek
Department of Computer Science and Engineering (AI & ML)
Haldia Institute of Technology
Haldia, West Bengal, India

Jabulani Christopher Nyawo
School of Management, IT and Governance
University of KwaZulu-Natal
Durban, South Africa

Chandan Paul
Electrical Engineering Department
Dr. B. C. Roy Engineering College
Durgapur, WB, India

Sourav Paul
Electrical Engineering Department
Dr. B. C. Roy Engineering College
Durgapur, WB, India

Dharmbir Prasad
Department of Electrical and Electronics Engineering
SRM Institute of Science & Technology
Delhi-NCR Campus, India

Pranati Rani Purohit
Department of Electrical Engineering
National Institute of Technology
Rourkela, Odisha, India

Prince Raj
Electrical Engineering Department
National Institute of Technology
Rourkela, Odisha, India

Chikesh Ranjan
Department of Electrical Engineering
National Institute of Technology
Rourkela, Odisha, India

Pravat Kumar Ray
Department of Electrical Engineering
National Institute of Technology
Rourkela, Odisha, India

Apurba Kumar Roy
Department of Electrical Engineering
Birla Institute of Technology
Mesra, Ranchi, India

Provas Kumar Roy
Department of Electrical Engineering
Kalyani Government Engineering College
Kalyani, WB, India

Ranadip Roy
Department of Electrical Engineering
Sanaka Educational Trust's Group of Institutions
Durgapur, West Bengal, India

Koushik Sarkar
Department of Electrical Engineering
Asansol Engineering College
Asansol, West Bengal, India

Tushnik Sarkar
Electrical Engineering Department
Dr. B. C. Roy Engineering College
Durgapur, WB, India

Sameer Sharma
Electrical Engineering Department
Dr. B. C. Roy Engineering College
Durgapur, WB, India

Rishabh Dev Shukla
Electrical Engineering Department
The Assam Kaziranga University
Koraikhowa, Assam, India

Rudra Pratap Singh
Asansol Engineering College
Asansol, West Bengal, India

Sneha Sultana
Electrical Engineering Department
Dr. B. C. Roy Engineering College
Durgapur, WB, India

Vilay Vannaladsaysy
Department of Mechanical Engineering
National University of Laos
Sokpaluang, Vientiane, Laos

Yusuf Olatunji Waidi
Department of Materials Engineering
Indian Institute of Science
Bangalore, India

Chapter 1

Renewable energy

Opportunities, applications, and future scope of artificial intelligence and internet of things

Dipanwita Ghosh, Arpan Adhikary, Asit Kumar Nayek, and Sunanda Hazra

INTRODUCTION

The global population is increasingly gravitating toward urban centers, with estimates suggesting that by 2050, nearly 70% of humans will reside in cities. This unprecedented urbanization poses profound challenges, including the exponential increase in energy consumption and its associated environmental impacts [1]. Concurrently, the pressing need to combat climate change has propelled the adoption of renewable energy (RE) sources as a cornerstone of sustainable development strategies worldwide [2]. In this context, the emergence of smart cities has provided a compelling framework for addressing the complex interplay between urbanization, energy demand, and environmental sustainability. Smart cities leverage advanced technologies and data-driven approaches to optimize resource utilization, enhance efficiency, and improve the quality of life for residents [3]. At the heart of this transformation lies the integration of RE solutions, which offer a myriad of benefits ranging from carbon emissions reduction to energy security and economic resilience. In an era marked by unprecedented urbanization and pressing environmental concerns, the concept of smart cities has emerged as a beacon of hope for sustainable urban development [4]. Smart cities leverage advanced technologies and data-driven approaches to optimize resource utilization, enhance quality of life, and promote economic prosperity. At the heart of this transformation lies the integration of RE sources, which offer a pathway toward a cleaner, more resilient urban future [5]. RE technologies, including solar, wind, hydroelectric, biomass, and geothermal power, present a compelling solution to the challenges of climate change and energy sustainability. By harnessing these abundant and inexhaustible resources, smart cities can reduce greenhouse gas emissions, enhance energy security, and drive innovation across various sectors [6]. This chapter aims to explore the critical role of RE integration in the context of smart city development. Through an in-depth analysis of theoretical frameworks, case studies, and practical

DOI: 10.1201/9781003628279-1

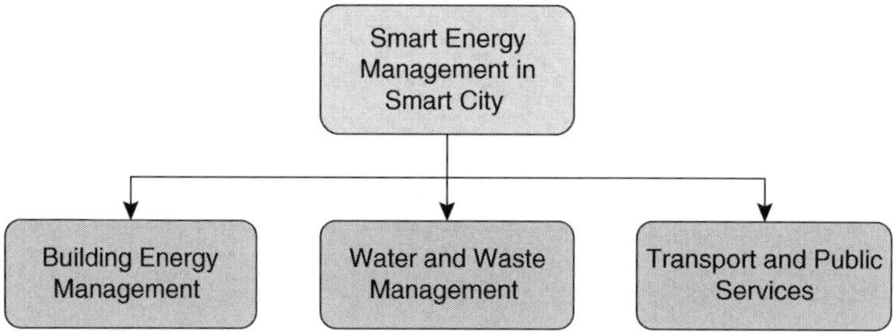

Figure 1.1 Smart energy management in smart cities.

applications, we will examine the benefits, challenges, and opportunities associated with leveraging RE in smart urban environments. Additionally, we will explore the synergies between RE deployment and other smart city initiatives, such as intelligent infrastructure, data analytics, and sustainable mobility solutions. By elucidating the transformative potential of RE in smart cities, this chapter seeks to provide policymakers, urban planners, and stakeholders with actionable insights and strategies for fostering sustainable urban development. Through collaborative efforts and innovative approaches, we can harness the power of RE to build smarter, greener, and more resilient cities for generations to come (Figure 1.1).

RENEWABLE ENERGY IN SMART CITY

As urbanization accelerates globally, the concept of smart cities has emerged as a strategic approach to address the challenges of sustainable urban development. Smart cities leverage advanced technologies, data analytics, and innovative solutions to optimize resource management, enhance efficiency, and improve the quality of life for residents [7]. At the core of this transformation lies the integration of RE sources, which offer a sustainable alternative to traditional fossil fuels and pave the way for a cleaner, greener urban future [8]. RE encompasses a diverse array of sources, including solar, wind, hydroelectric, biomass, and geothermal power [9]. These resources hold immense potential to meet the energy needs of urban populations while mitigating the environmental impacts associated with conventional energy generation. In the context of smart cities, the deployment of RE technologies presents a multitude of opportunities and benefits as stated below:

Sustainability

RE sources generate electricity without depleting finite resources or emitting harmful greenhouse gases, thereby reducing carbon emissions and mitigating the adverse effects of climate change. By transitioning to renewables, smart cities can contribute to global efforts to achieve carbon neutrality and promote environmental sustainability.

Energy security

Diversifying the energy mix with renewable sources enhances energy security by reducing dependence on imported fossil fuels and mitigating the risks associated with supply disruptions. Smart cities can leverage local renewable resources to enhance resilience against external shocks and ensure a reliable and resilient energy supply for residents and businesses [10].

Economic development

The deployment of RE infrastructure stimulates economic growth, fosters innovation, and creates job opportunities across various sectors. Smart cities that invest in RE technologies attract investment, spur entrepreneurship, and enhance competitiveness in the global marketplace, contributing to long-term economic prosperity [11].

Technological innovation

RE integration drives innovation in energy storage, grid management, and digitalization, enabling smarter and more efficient energy systems. By leveraging advanced technologies such as IoT, AI, and blockchain, smart cities can optimize energy production, distribution, and consumption, leading to greater efficiency and sustainability.

Resilience

RE systems enhance the resilience of urban infrastructure by enabling decentralized energy generation and microgrid solutions. In the face of natural disasters or grid failures, smart cities equipped with RE sources can maintain essential services, ensure continuity of operations, and enhance community resilience, thereby safeguarding the well-being of residents [12].

In the pursuit of sustainable urban development, the integration of smart transport systems has become a cornerstone of modern city planning. Smart cities leverage advanced technologies and data-driven solutions to address challenges in urban mobility, reduce congestion, enhance efficiency, and promote environmental sustainability. Smart transport encompasses a diverse array of strategies and innovations aimed at revolutionizing how people and goods move within urban environments [13]. This comprehensive overview delves into the key components, benefits, and challenges of smart transport in the context of smart cities.

KEY COMPONENTS OF SMART TRANSPORT

Intelligent transportation system (ITS)

ITS integrates information and communication technologies to optimize the efficiency and safety of transportation networks. This includes real-time traffic monitoring, dynamic route planning, and smart traffic signal control systems to alleviate congestion and improve traffic flow.

Multimodal mobility solutions

Smart cities embrace multimodal transportation options, including public transit, cycling infrastructure, ride-sharing services, and pedestrian-friendly urban design. By providing diverse and interconnected transportation choices, cities can reduce reliance on private vehicles and promote sustainable travel behavior [14].

Electric and autonomous vehicles

The adoption of electric vehicles (EVs) and autonomous vehicles (AVs) represents a paradigm shift in urban mobility. EVs offer a cleaner alternative to traditional gasoline-powered cars, reducing air pollution and greenhouse gas emissions. AVs have the potential to enhance road safety, reduce traffic accidents, and optimize traffic flow through advanced automation and vehicle-to-infrastructure communication [15].

Data analytics and mobility platforms

Data-driven approaches play a crucial role in optimizing transport systems and improving user experience. Mobility platforms and apps provide real-time information on transit schedules, traffic conditions, parking availability, and alternative travel options, empowering users to make informed decisions and navigate the city more efficiently [16].

BENEFITS OF SMART TRANSPORT IN SMART CITIES

Reduced congestion and emissions

Smart transport systems help alleviate traffic congestion and reduce air pollution by optimizing traffic flow, promoting alternative modes of transportation, and encouraging the adoption of clean vehicles.

Enhanced efficiency and accessibility

By providing seamless connectivity and real-time information, smart transport improves the accessibility of urban areas, enhances the efficiency of transportation networks, and reduces travel times for residents and commuters.

Improved safety and resilience

Intelligent transportation systems enhance road safety through features such as automated collision avoidance, adaptive cruise control, and emergency response coordination [17]. Additionally, smart transport infrastructure enhances the resilience of cities by mitigating the impacts of extreme weather events and natural disasters on transportation networks.

In the digital age, communication lies at the heart of every thriving community, and in the context of smart cities, it becomes the backbone of efficient urban living. Smart communication encompasses a spectrum of technologies, platforms, and strategies aimed at enhancing connectivity, facilitating collaboration, and empowering residents and businesses to actively participate in city life. This comprehensive overview delves into the key components, benefits, and challenges of smart communication in the context of smart cities.

KEY COMPONENTS OF SMART COMMUNICATION

Fiber optic networks and 5G infrastructures

High-speed internet connectivity forms the foundation of smart communication, enabling seamless data transmission, real-time information exchange, and IoT connectivity. Fiber optic networks and 5G infrastructure provide the bandwidth and low latency required to support a wide range of smart city applications, from smart traffic management to remote healthcare services [18].

IoT and sensor networks

IoT devices and sensor networks play a crucial role in gathering real-time data on various aspects of urban life, including traffic flow, air quality, energy consumption, and waste management. These data streams enable city officials to make informed decisions, optimize resource allocation, and improve the overall quality of life for residents.

Smart grids and energy management systems

Smart communication facilitates the integration of RE sources, energy storage systems, and demand-response mechanisms into the electrical grid. By enabling bidirectional communication between energy producers, consumers, and grid operators, smart grids optimize energy distribution, reduce wastage, and enhance grid resilience [19].

Digital platforms and citizen engagement tools

Digital platforms and mobile applications empower residents to engage with city services, report issues, and participate in decision-making processes. From e-government portals to crowdsourcing platforms, smart communication tools foster transparency, accountability, and civic participation, strengthening the social fabric of smart cities [20].

BENEFITS OF SMART COMMUNICATION IN SMART CITIES

Enhanced connectivity and accessibility

Smart communication technologies bridge the digital divide, providing ubiquitous connectivity and ensuring equal access to information and services for all residents, regardless of socioeconomic status or geographic location.

Improved efficiency and resource management

Real-time data exchange and analytics enable city officials to optimize resource allocation, streamline operations, and respond swiftly to emerging challenges, such as traffic congestion, air pollution, and public safety incidents [21].

Empowered communities and citizen-centric services

Smart communication tools empower residents to actively participate in city governance, voice their concerns, and contribute to community initiatives. By fostering transparency, collaboration, and co-creation, smart cities become more inclusive, resilient, and responsive to the needs of their inhabitants.

Innovation and economic development

The deployment of smart communication infrastructure fosters innovation ecosystems, attracting talent, investment, and entrepreneurship to the city. By nurturing a culture of innovation and collaboration, smart cities stimulate economic growth, create job opportunities, and enhance competitiveness in the global marketplace [22].

In the rapidly evolving landscape of urban living, the convergence of healthcare and technology has sparked a revolution in the way cities approach public health and wellness. Smart healthcare, an integral component of the broader concept of smart cities, leverages advanced technologies, data analytics, and innovative solutions to enhance accessibility, efficiency, and quality of healthcare services [23]. This comprehensive overview explores the key components, benefits, and challenges of smart healthcare in the context of smart cities.

KEY COMPONENTS OF SMART HEALTHCARE

Telemedicine and remote monitoring

Telemedicine platforms enable remote consultations, diagnosis, and treatment delivery, facilitating access to healthcare services for residents regardless of geographic location. Remote monitoring devices, such as wearables and IoT-enabled sensors, allow healthcare providers to monitor patients' vital signs, chronic conditions, and medication adherence remotely, enabling proactive intervention and personalized care [24].

Health information systems and electronic health records (EHRs)

Digital health information systems and EHRs streamline medical records management, enabling seamless data exchange and interoperability among healthcare providers. Centralized health databases and electronic medical records enhance care coordination, reduce administrative burden, and improve patient outcomes through data-driven decision-making and predictive analytics.

IoT-enabled healthcare devices and wearables

IoT-enabled healthcare devices, such as smart wearables, fitness trackers, and connected medical devices, empower individuals to monitor their health status, track fitness goals, and manage chronic conditions in real time. These devices provide valuable insights into individuals' health behaviors, enabling early detection of health issues and proactive intervention to prevent disease progression [25].

Health analytics and predictive modeling

Advanced analytics and predictive modeling techniques leverage healthcare data to identify trends, predict disease outbreaks, and optimize resource allocation. By analyzing population health data, healthcare providers and policymakers can tailor interventions, allocate resources efficiently, and implement targeted health promotion initiatives to improve public health outcomes [26].

BENEFITS OF SMART HEALTHCARE IN SMART CITIES

Enhanced access and convenience

Smart healthcare technologies break down barriers to healthcare access by providing remote consultations, telemedicine services, and digital health platforms that enable residents to access healthcare services anytime, anywhere.

Improved patient outcomes and quality of care

By leveraging data analytics and remote monitoring solutions, smart healthcare enables proactive intervention, personalized treatment plans, and continuous monitoring of patients' health status, leading to improved health outcomes and enhanced quality of care [27].

Efficient resource utilization and cost savings

Smart healthcare systems optimize resource allocation, reduce unnecessary hospital visits, and prevent avoidable readmissions through remote monitoring, predictive analytics, and early intervention, resulting in cost savings for healthcare providers and payers.

Empowered patients and health promotion

Smart healthcare empowers individuals to take control of their health through self-monitoring, health education, and personalized wellness programs. By promoting health literacy and encouraging preventive behaviors, smart healthcare fosters a culture of wellness and resilience in urban communities.

RELATIONSHIP BETWEEN ARTIFICIAL INTELLIGENCE AND RENEWABLE ENERGY

RE can provide a long-term solution to the needed energy. Climate change, increasing pollution, and resource diminution are the main reasons behind the increasing potential of RE. As RE is mainly available in remote areas, this becomes a challenging development and growth of technology. The main advantage of RE is that it is independent of fossil fuels, but the energy sources face different fluctuations. This fluctuation of RE forces us to adopt emerging technologies toward storing and producing energy in favorable conditions and use it while the condition is difficult. AI considered RE for studying and different calculations that allow the perception, reasoning, and necessary actions according to need. As per the experts [28], AI can recommend, take decisions, and predict human-defined plans and actions in a real or virtual environment [29]. As per the grid's growing complexity, AI can apply different automated algorithms on climate prediction to maximize the return on the investment in RE and estimate the power consumption of all the grid's production components. AI can also be used to collect and evaluate a heavy amount of data from all the key features of a grid and also to provide rapid classification for the optimal resource distribution in and out of the grid. For better operation and maintenance, the following AI applications need to be focused:

- RE applied energy production.
- Operational stability, reliability, and security.
- Demand forecasting, identification, and accuracy.
- Optimal storage and energy generation.

IoT-enabled devices are useful in collecting data and storing them to the cloud storage instantly. IoT-enabled devices can be useful in finding the RE resources and detecting the area covered by them. Also, IoT devices can be useful to find the type of energy and other important insights from the energy sources. Being internet-enabled, these devices are capable of sending the data to other nodes as well as to the storage servers.

CONCLUSION

In conclusion, RE integration within the framework of smart cities represents a pivotal step toward achieving sustainable urban development. By harnessing the power of renewable resources such as solar, wind, hydroelectric, biomass, and geothermal energy, smart cities can address pressing environmental challenges while enhancing energy security, promoting economic prosperity, and improving the overall quality of life for residents

[30]. The adoption of RE technologies in smart cities offers a multitude of benefits, including the reduction of carbon emissions, mitigation of climate change impacts, diversification of energy sources, and creation of green jobs. Moreover, by embracing RE solutions, smart cities can drive innovation, foster technological advancement, and attract investment, positioning themselves as leaders in the global transition to clean energy. However, the successful integration of RE into smart city initiatives requires overcoming various challenges, including regulatory barriers, financial constraints, technological limitations, and public acceptance. Collaboration among governments, businesses, academia, and communities is essential to address these challenges and unlock the full potential of RE in shaping sustainable urban futures. As we move forward, it is imperative for policymakers, urban planners, and stakeholders to prioritize RE integration in smart city planning and implementation efforts. By adopting holistic approaches, leveraging innovative technologies, and fostering inclusive decision-making processes, smart cities can pave the way for a more resilient, equitable, and environmentally sustainable urban future. In doing so, they will not only enhance the well-being of their residents but also contribute to global efforts to combat climate change and build a more sustainable planet for generations to come.

REFERENCES

1. Zhang, L., Ling, J., & Lin, M. (2022). Artificial intelligence in renewable energy: A comprehensive bibliometric analysis. *Energy Reports*, *8*, 14072–14088.
2. Abdalla, A. N., Nazir, M. S., Tao, H., Cao, S., Ji, R., Jiang, M., & Yao, L. (2021). Integration of energy storage system and renewable energy sources based on artificial intelligence: An overview. *Journal of Energy Storage*, *40*, 102811.
3. Chen, C., Hu, Y., Karuppiah, M., & Kumar, P. M. (2021). Artificial intelligence on economic evaluation of energy efficiency and renewable energy technologies. *Sustainable Energy Technologies and Assessments*, *47*, 101358.
4. Şerban, A. C., & Lytras, M. D. (2020). Artificial intelligence for smart renewable energy sector in Europe: Smart energy infrastructures for next generation smart cities. *IEEE Access*, *8*, 77364–77377.
5. Hannan, M. A., Al-Shetwi, A. Q., Ker, P. J., Begum, R. A., Mansor, M., Rahman, S. A., & Muttaqi, K. M. (2021). Impact of renewable energy utilization and artificial intelligence in achieving sustainable development goals. *Energy Reports*, *7*, 5359–5373.
6. Boza, P., & Evgeniou, T. (2021). Artificial intelligence to support the integration of variable renewable energy sources to the power system. *Applied Energy*, *290*, 116754.

7. Al-Othman, A., Tawalbeh, M., Martis, R., Dhou, S., Orhan, M., Qasim, M., & Olabi, A. G. (2022). Artificial intelligence and numerical models in hybrid renewable energy systems with fuel cells: Advances and prospects. *Energy Conversion and Management*, *253*, 115154.
8. He, Q., Zheng, H., Ma, X., Wang, L., Kong, H., & Zhu, Z. (2022). Artificial intelligence application in a renewable energy-driven desalination system: A critical review. *Energy and AI*, *7*, 100123.
9. Ahmad, T., Zhang, D., Huang, C., Zhang, H., Dai, N., Song, Y., & Chen, H. (2021). Artificial intelligence in sustainable energy industry: Status Quo, challenges and opportunities. *Journal of Cleaner Production*, *289*, 125834.
10. Kanase-Patil, A. B., Kaldate, A. P., Lokhande, S. D., Panchal, H., Suresh, M., & Priya, V. (2020). A review of artificial intelligence-based optimization techniques for the sizing of integrated renewable energy systems in smart cities. *Environmental Technology Reviews*, *9*(1), 111–136.
11. Mazzeo, D., Herdem, M. S., Matera, N., Bonini, M., Wen, J. Z., Nathwani, J., & Oliveti, G. (2021). Artificial intelligence application for the performance prediction of a clean energy community. *Energy*, *232*, 120999.
12. Rangel-Martinez, D., Nigam, K. D. P., & Ricardez-Sandoval, L. A. (2021). Machine learning on sustainable energy: A review and outlook on renewable energy systems, catalysis, smart grid and energy storage. *Chemical Engineering Research and Design*, *174*, 414–441.
13. Hassan, Q., Sameen, A. Z., Salman, H. M., Al-Jiboory, A. K., & Jaszczur, M. (2023). *The role of renewable energy and artificial intelligence towards environmental sustainability and net zero*. Research Square, Springer, 1–25. https://doi.org/10.21203/rs.3.rs-2970234/v1
14. Garcia Marquez, F. P., & Peinado Gonzalo, A. (2022). A comprehensive review of artificial intelligence and wind energy. *Archives of Computational Methods in Engineering*, *29*(5), 2935–2958.
15. Liu, Z., Sun, Y., Xing, C., Liu, J., He, Y., Zhou, Y., & Zhang, G. (2022). Artificial intelligence powered large-scale renewable integrations in multi-energy systems for carbon neutrality transition: Challenges and future perspectives. *Energy and AI*, *10*, 100195.
16. Entezari, A., Aslani, A., Zahedi, R., & Noorollahi, Y. (2023). Artificial intelligence and machine learning in energy systems: A bibliographic perspective. *Energy Strategy Reviews*, *45*, 101017.
17. Mohammad, A., & Mahjabeen, F. (2023). Revolutionizing solar energy: The impact of artificial intelligence on photovoltaic systems. *International Journal of Multidisciplinary Sciences and Arts*, *2*(1), 117–127.
18. Lai, J. P., Chang, Y. M., Chen, C. H., & Pai, P. F. (2020). A survey of machine learning models in renewable energy predictions. *Applied Sciences*, *10*(17), 5975.
19. Ali, S. S., & Choi, B. J. (2020). State-of-the-art artificial intelligence techniques for distributed smart grids: A review. *Electronics*, *9*(6), 1030.
20. Albarakati, A. J., Boujoudar, Y., Azeroual, M., Jabeur, R., Aljarbouh, A., El Moussaoui, H., & Ouaaline, N. (2021). Real-time energy management for DC microgrids using artificial intelligence. *Energies*, *14*(17), 5307.

21. Abualigah, L., Zitar, R. A., Almotairi, K. H., Hussein, A. M., Abd Elaziz, M., Nikoo, M. R., & Gandomi, A. H. (2022). Wind, solar, and photovoltaic renewable energy systems with and without energy storage optimization: A survey of advanced machine learning and deep learning techniques. *Energies*, *15*(2), 578.
22. Saheb, T., Dehghani, M., & Saheb, T. (2022). Artificial intelligence for sustainable energy: A contextual topic modeling and content analysis. *Sustainable Computing: Informatics and Systems*, *35*, 100699.
23. Wang, H., Liu, Y., Zhou, B., Li, C., Cao, G., Voropai, N., & Barakhtenko, E. (2020). Taxonomy research of artificial intelligence for deterministic solar power forecasting. *Energy Conversion and Management*, *214*, 112909.
24. Yousuf, H., Zainal, A. Y., Alshurideh, M., & Salloum, S. A. (2020). Artificial intelligence models in power system analysis. In: Aboul Ella Hassanien, Roheet Bhatnagar, Ashraf Darwish (eds), *Artificial Intelligence for Sustainable Development: Theory, Practice and Future Applications* (pp. 231–242). Cham: Springer International Publishing.
25. Li, J., Herdem, M. S., Nathwani, J., & Wen, J. Z. (2023). Methods and applications for artificial intelligence, big data, internet of things, and blockchain in smart energy management. *Energy and AI*, *11*, 100208.
26. Mohammadi, E., Alizadeh, M., Asgarimoghaddam, M., Wang, X., & Simões, M. G. (2022). A review on application of artificial intelligence techniques in microgrids. *IEEE Journal of Emerging and Selected Topics in Industrial Electronics*, *3*(4), 878–890.
27. Sabzehgar, R., Amirhosseini, D. Z., & Rasouli, M. (2020). Solar power forecast for a residential smart microgrid based on numerical weather predictions using artificial intelligence methods. *Journal of Building Engineering*, *32*, 101629.
28. Shi, C., Feng, X., & Jin, Z. (2022). Sustainable development of China's smart energy industry based on artificial intelligence and low-carbon economy. *Energy Science & Engineering*, *10*(1), 243–252.
29. Thirunavukkarasu, M., Sawle, Y., & Lala, H. (2023). A comprehensive review on optimization of hybrid renewable energy systems using various optimization techniques. *Renewable and Sustainable Energy Reviews*, *176*, 113192.
30. Huang, J., & Koroteev, D. D. (2021). Artificial intelligence for planning of energy and waste management. *Sustainable Energy Technologies and Assessments*, *47*, 101426.

Chapter 2

Machine learning-enhanced maximum power point tracking for solar modules in photovoltaic systems

Prince Raj, Prasenjit Dey, and Arnab Ghosh

Renewable energy is becoming increasingly popular due to its abundance and clean nature. Solar, wind, geothermal, and biomass are the main types of renewable energy [1]. Solar energy can be captured as heat or converted to electricity [2]. Wind turbines use wind speed to generate electricity. Geothermal power plants use the heat from the Earth's core to generate electricity. Bioenergy uses plant material to generate electricity. Many countries are setting ambitious goals to increase their use of renewable energy.

On top of that, solar energy plays a vital role in renewable energy sources because sun rays reach almost everywhere on the Earth, making it the most abundant form of energy in nature. So, solar panels can be easily installed on houses, businesses, or large utility-scale farms [3–5]. It ultimately helps reduce greenhouse gases, as fossil fuels are not consumed for solar energy production, which is crucial in protecting humans, wildlife, and the ecosystem. In the past few decades, engineers have developed several algorithms to implement maximum power point tracking to design solar inverters that can extract maximum power generated by photovoltaic systems [6–8]. Factors like solar irradiance and temperature are accounted for to ensure that the system generates maximum power. In practical situations, neither solar irradiance nor temperatures received by the solar panels are consistent throughout the day. This inconsistency leads to variable power generation by the solar panels, and it becomes impossible to meet the power requirements of the load wherever it is installed and connected to [9–12].

As solar irradiance increases throughout the day, the number of photons striking the per unit surface area of the PV panels increases, resulting in increased power production and, thus, increased efficiency of the overall system. Solar panels work best in cooler temperatures. Solar panels require semiconductors to convert sunlight into electricity. As heat increases, these materials become more agitated, lowering their ability to efficiently convert light into electricity. When a panel becomes hotter, there is a more significant drop in voltage output. Voltage is like water pressure in electricity; less power is generated due to reduced voltage output. Another reason solar panels become less effective under high-temperature conditions is due to

DOI: 10.1201/9781003628279-2

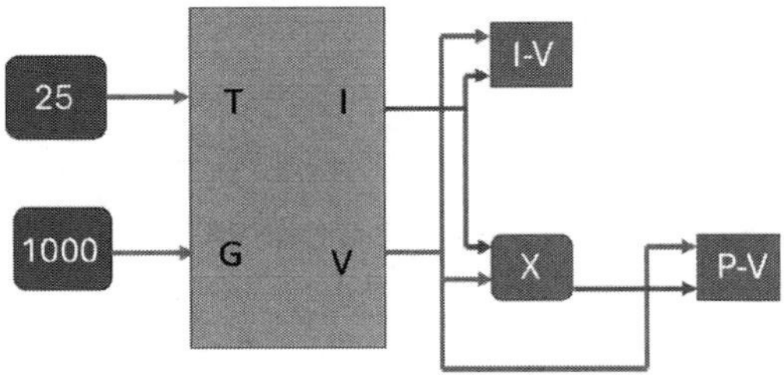

Figure 2.1 Schematic diagram of the MPPT circuit implemented for the study.

increased electrical leakage within the panel, which subsequently lowers overall performance [13]. There are lots of conventional algorithms to track the maximum power, like perturbation and observation (P&O) [14–18], fractional open-circuit voltage [19], and incremental conductance [20]. A proper dataset that contains the data about solar irradiance, temperature, voltage at MPP (maximum power point), and current at MPP is used for training the data models using machine learning (ML) algorithms, namely gradient descent and neural network [21–25]. These ML-based models have been proposed to solve the same old-era problems to track MPP, but the main challenge is to train the models properly on the reference dataset. Figure 2.1 shows a schematic diagram of the MPP circuit implemented for the study of power generation by solar panels. The solar panel takes two inputs, temperature (T) and solar irradiance (G), and outputs, current and voltage, respectively. These outputs are used to generate I–V and P–V curves of the solar panel for better visualization and speculation of the system's MPP.

In this study, we are considering all the scenarios at a variable temperature and variable solar irradiance (G) that is created by [26–30]. The temperature value varies from 15°C to 40°C in steps of 1°C, and the value of solar irradiance varies between 200 W/m^2 and 1090 W/m^2. At the specific conditions, we wish to implement a model using the concepts of ML that can predict the value of output current through solar panels at MPP, which could further be used to find out the maximum output power produced by the PV panel. So, first, we will use gradient descent to implement a model to predict the output current at MPP, and then, we will use multilayer perception. So, to begin with gradient descent, we will consider a hypothesis in which we will consider only one input, i.e., varying solar irradiance (G), and the value of predicted output current will vary linearly with the given input.

Let the hypothesis be given as

Hypothesis: $h_\theta(x) = \theta_0 + \theta_1 x$
Parameters: θ_0, θ_1

where x represents the training input data samples, $h_\theta(x)$ represents the predicted output at input training data samples; θ_0, θ_1 denote the parameters that are needed to be tuned for efficient working of the model. Let y denote the actual output of the training data samples. So, our error function [14] will be

$$J(\theta_0,\theta_1)=\frac{1}{2m}\sum_{i=1}^{m}\left(h_\theta\left(x^{(i)}\right)-\left(y^{(i)}\right)\right)^2 \tag{2.1}$$

where m represents the total number of training examples.

In gradient descent, we are trying to find the optimal value of the parameters, θ_0, θ_1, to minimize the error function $J(\theta_0, \theta_1)$.

To update the parameters $J(\theta_0, \theta_1)$, we use the equation

$$\theta_j := \theta_j - \alpha\frac{\partial}{\partial\theta_j}J(\theta_0,\theta_1)\quad(\text{for } j=0,1) \tag{2.2}$$

where α represents the learning rate, which determines the size of each step in the direction of minimizing the loss function.

Now, we introduce one hidden layer consisting of 20 neurons for a multilayer perceptron (MLP). Each neuron has a bias, and each edge consists of a weight. A general model of the same is depicted in Figure 2.2, where T represents temperature and G represents irradiance. We assume the initial values of all the weights and biases to be any random value between –1 and +1. Then, we perform forward propagation on the given neural network (Figure 2.2).

So, first, we will obtain the output at the hidden layers, which will be the product of the respective edge weight and the bias of the neuron.

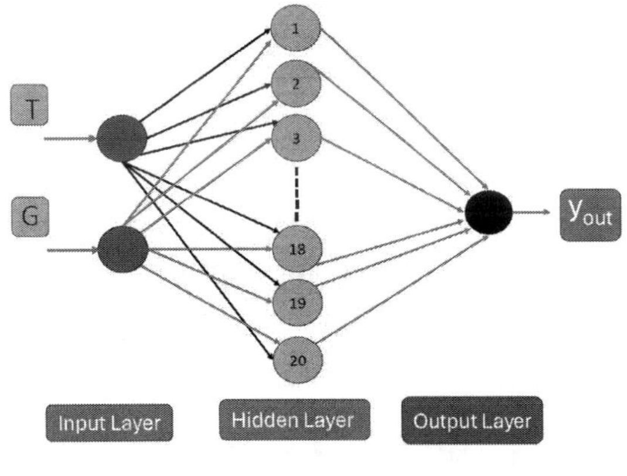

Figure 2.2 Neural network block diagram for one hidden layer.

$$y_1 = w_{T1} \cdot T_1 + w_{G1} \cdot G_1 + b_1$$
$$y_2 = w_{T2} \cdot T_2 + w_{G2} \cdot G_2 + b_2$$
$$\vdots \tag{2.3}$$
$$y_{20} = w_{T20} \cdot T_{20} + w_{G20} \cdot G_{20} + b_{20}$$

where x is the input given to the neural network.

$$y_{\text{out}} = \sum_{i=1}^{20} y_i \cdot w_i^{(2)} + b_{\text{out}} \tag{2.4}$$

It is interesting to note that the activation function used here is the rectified linear activation function, given mathematically as

$$f(x) = \max(0, x) \tag{2.5}$$

After applying the first forward propagation pass, we must tune all the weights and biases so they can perform well on unseen data. We will use the concept of backpropagation for that. We'll allow all the training data to loop through one backward pass to adjust all the weights and biases [18]. The exact process, starting from one pass of forward propagation, is repeated with one pass of backward propagation, along with updating the values of weights and biases on all available training examples [19]. This process is repeated for several iterations to adjust the model parameters well.

Now, updated weights and biases can be obtained by the following formulas

$$W_{\text{new}} = W_{\text{old}} + \delta W \tag{2.6}$$

$$\theta_{\text{new}} = \theta_{\text{old}} + \delta\theta \tag{2.7}$$

where δW and $\delta\theta$ represent change in weight and bias during backward propagation, respectively [20].

Table 2.1 represents the part of the dataset that is used for training the data models to predict the current at MPP. It is to be noted here that this table corresponds to variable temperature and variable solar irradiance.

Tables 2.2 and 2.3 show the final values of weights of inputs in the hidden layer and weights in the output layer, respectively, obtained after all the 1000 iterations over all the training examples. Tables 2.4 and 2.5 show the final values of bias of inputs in hidden layer and bias in output layer, respectively, obtained after all the 1000 iterations over all the training examples.

Table 2.1 Training data for MPP prediction

	Sample dataset used		
Serial	*Temperature (°C)*	*Solar irradiance (W/M²)*	*Current at MPP(A)*
1	15	200	1.4738
2	16	240	1.7814
3	18	600	4.628
4	25	1000	7.5764
5	40	1000	7.445

Table 2.2 Weights of input in the hidden layer

Serial	*T*	*G*
1	0.04521137	0.79307225
2	0.40902416	0.5047629
3	0.26116629	0.23229665
4	−0.04458023	−0.33671209
5	−0.10933988	−0.4648656
6	0.35013229	0.51250282
7	−0.20011079	−1.23183563
8	0.84998239	−0.11723376
9	0.87906987	−0.28857376
10	−0.26854495	−0.4146936
11	0.60169752	−0.34520578
12	0.03781908	0.45305626
13	0.28905702	0.52231008
14	0.74615629	−0.5859159
15	−0.93360453	−1.48365866
16	−0.91682496	−0.38986879
17	−0.91431007	0.44710944
18	0.59719428	−0.23038576
19	0.54160631	0.85461584
20	0.65914874	−0.15959109

After calculating the mean absolute error, it is clear that the MLP with 20 neurons is quite as compared to MLP with 15 neurons. The number of epochs for both gradient descent and neural network is 1000, and the learning rates for both gradient descent and neural network are 1e−8 and 1e−6, respectively. Figures 2.3–2.5 show the comparison of actual and predicted current for a PV panel under different temperature and solar irradiance

Table 2.3 Weights in the output layer

Serial	*Weights in the output layer*
1	–0.58309698
2	–0.20904842
3	0.35231785
4	–1.03906454
5	0.33353343
6	0.36056455
7	–0.57923488
8	–4.34492978
9	–0.37129766
10	–0.27257846
11	0.14039354
12	–0.18897956
13	0.69100873
14	–0.80830163
15	–0.58224649
16	–0.70125015
17	0.28999162
18	–0.52443003
19	–0.11222309
20	–0.66243735

Table 2.4 Bias values for input in the hidden layer

Serial	*Bias values for input in the hidden layer*
1	–0.67934146
2	–0.77836746
3	0.3119583
4	–0.72042073
5	–0.60785205
6	–0.26321983
7	0.64375225
8	–0.76974257
9	0.67702192
10	–0.80697223
11	0.95249094
12	–0.06193201
13	0.95386674
14	0.21215652

(Continued)

Table 2.4 (Continued) Bias values for input in the hidden layer

Serial	*Bias values for input in the hidden layer*
15	0.48030213
16	–0.91948266
17	–0.43493048
18	–0.75800058
19	–0.40736789
20	–0.76029214

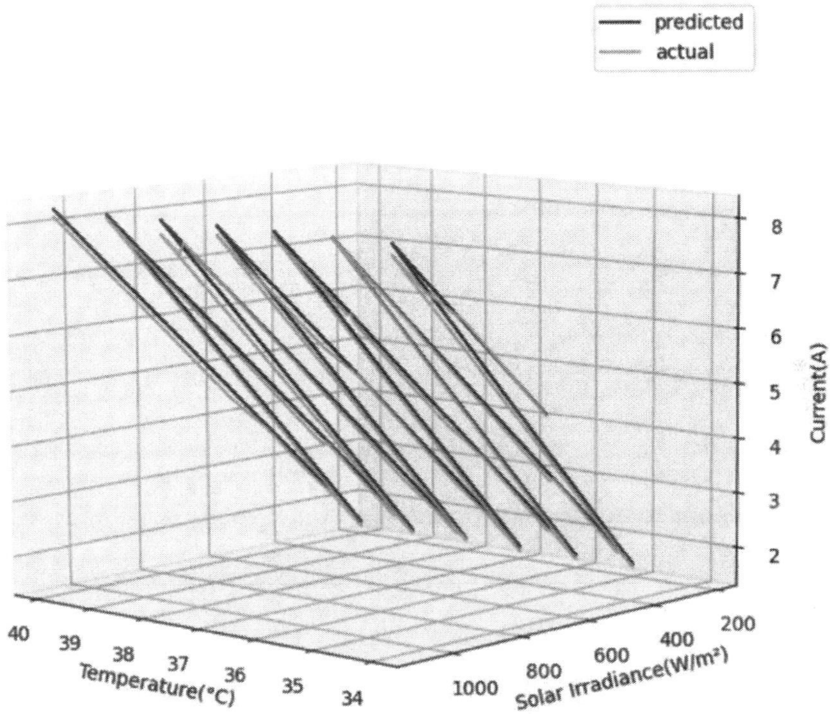

Figure 2.3 Comparison of actual and predicted current for a photovoltaic panel under different temperature and solar irradiance conditions using gradient descent.

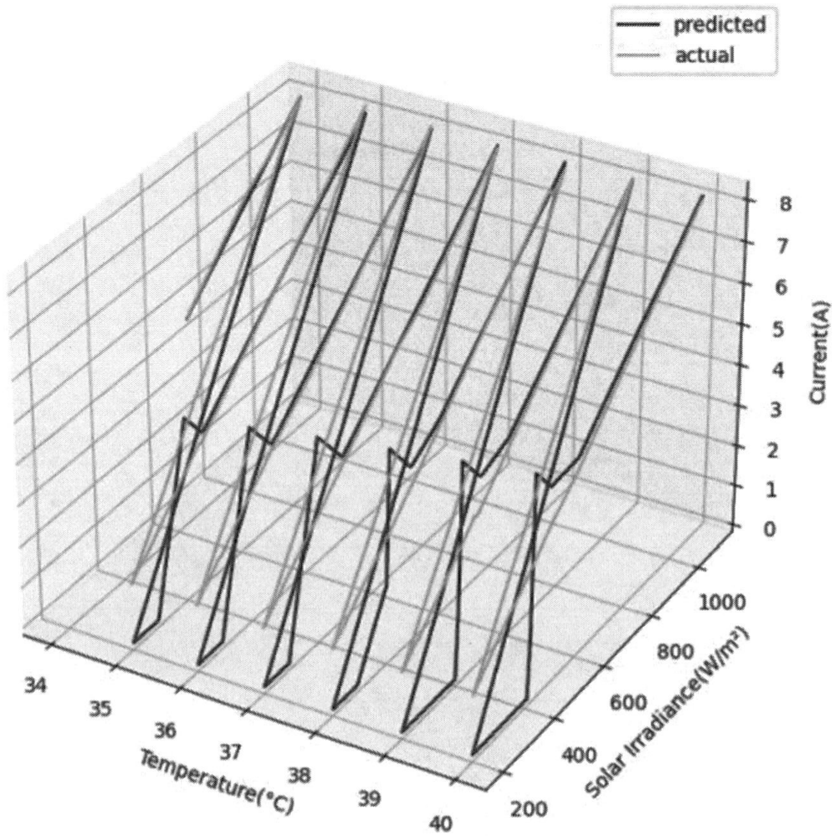

Figure 2.4 Comparison of actual and predicted current for a photovoltaic panel under different temperature and solar irradiance conditions using a multilayer perceptron consisting of 15 neurons.

conditions using gradient descent, MLP consisting of 20 neurons, and MLP composed of 15 neurons on testing data. If we try to compare these figures, the difference between actual and predicted values of currents is lesser in the case of MLP with 20 neurons than MLP with 15 neurons.

Figure 2.6 shows a comparative study between iterations versus the overall cost/error function of the given iteration for gradient descent and MLP with 20 hidden neurons and MLP with 15 hidden neurons. It is also evident that MLP with 20 neurons converges faster than gradient descent and MLP with 15 neurons. All these figures point in the direction that the efficiency of MLP is better compared to another algorithm for real-world complex problems. This algorithm learns by uncovering the hidden pattern from the data samples rather than relying on some random and rigid rules. This flexibility allows it to adapt better to variations in temperature and solar irradiance.

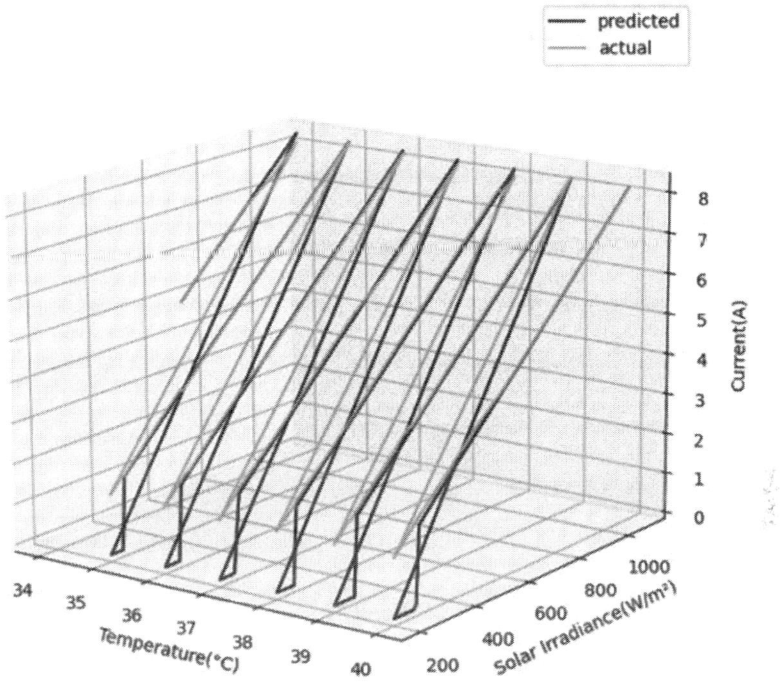

Figure 2.5 Comparison of actual and predicted current for a photovoltaic panel under different temperature and solar irradiance conditions using a multilayer perceptron consisting of 20 neurons.

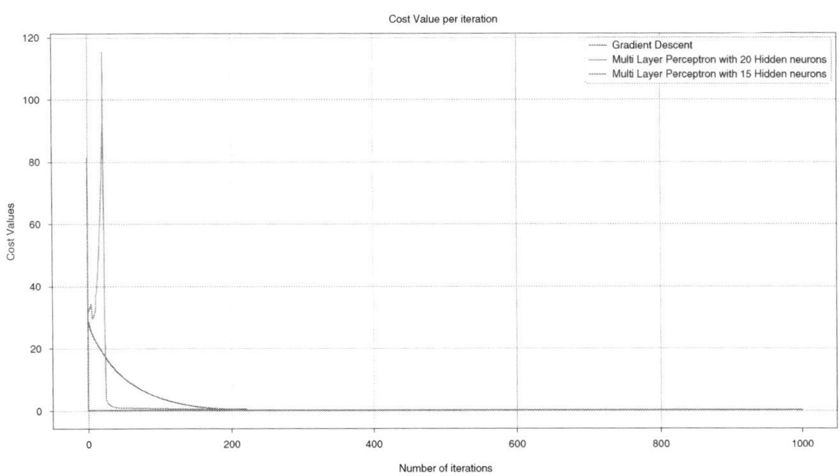

Figure 2.6 Comparative study between iterations versus overall cost/error function for the given iteration for gradient descent and multilayer perceptron with 20 hidden neurons and multilayer perceptron with 15 hidden neurons.

```
import numpy as np
import matplotlib.pyplot as plt
from mpl_toolkits import mplot3d

np.random.seed(0)

# Specify M: Number of training examples (75% of total data
    points)
# while 25% used for testing purpose
data = np.loadtxt("filename")

# Split data into features such as variable temperature,
# variable solar irradiance, and current at maximum power
    point.
y_train = np.array([])
input_assigned = np.array([])  # 2D array consisting of the
    two specified inputs for training data
x1_test = np.array([])  # Temperature of testing data
x2_test = np.array([])  # Solar Irradiance of testing data
y_test = np.array([])  # Current at MPP of testing data
input_assigned_test = np.array([])  # 2D array consisting of
     the two specified inputs for testing data

# Specify Learning rate
# Specify number of iterations
def backpropagation():
    # Update weights of inputs in hidden layer
    # Update weights of output layer
    # Update bias values of inputs in hidden layer
    # Update bias value of output layer
    return updated_weights, biases

# Specify the number of neurons, nneurons
# Specify the number of inputs, ninputs
weights1 = 2 * np.random.rand(ninputs, nneurons) - 1
weights2 = 2 * np.random.rand(nneurons, 1) - 1
bias1 = 2 * np.random.rand(nneurons, 1) - 1
bias2 = 2 * np.random.rand(1, 1) - 1
# (Initial values of all weights and biases are between -1 &
     +1)

for j in range(number_of_iterations):
    for i in range(M):
        # Perform forward propagation considering
        # Rectified linear activation function (ReLU)
        pass

error = np.array([])  # For calculating individual error for
     each and every testing data

for j in range(len(testing_data)):
    # Perform forward propagation considering
    # Rectified linear activation function (ReLU)
    error = np.append(error, actual_value - predicted_value)

# Calculate mean absolute error
# Plot error as a form of histogram for better visualization
fig = plt.figure()
ax = plt.axes(projection='3d')
ax.plot3D(x1_test_pred, x2_test_pred, y_test_pred, label="
    predicted")
ax.plot3D(x1_test, x2_test, y_test, label="actual")
ax.set_xlabel("Temperature")
ax.set_ylabel("Solar Irradiance")
ax.set_zlabel("Current")
ax.set_title('Label name')
plt.legend()
plt.grid()
plt.show()
```

The plot of histograms in Figures 2.7–2.9 represents the range of particular values of errors obtained over all the testing examples and the number of instances corresponding to that specific range for gradient descent, MLP with 15 neurons, and MLP with 20 neurons. With the help of obtained plots, it is easy to observe that in Figure 2.9, many testing examples have their errors near the zero error area, signifying that it has better predicting capabilities when compared to algorithms depicted in Figures 2.7 and 2.8.

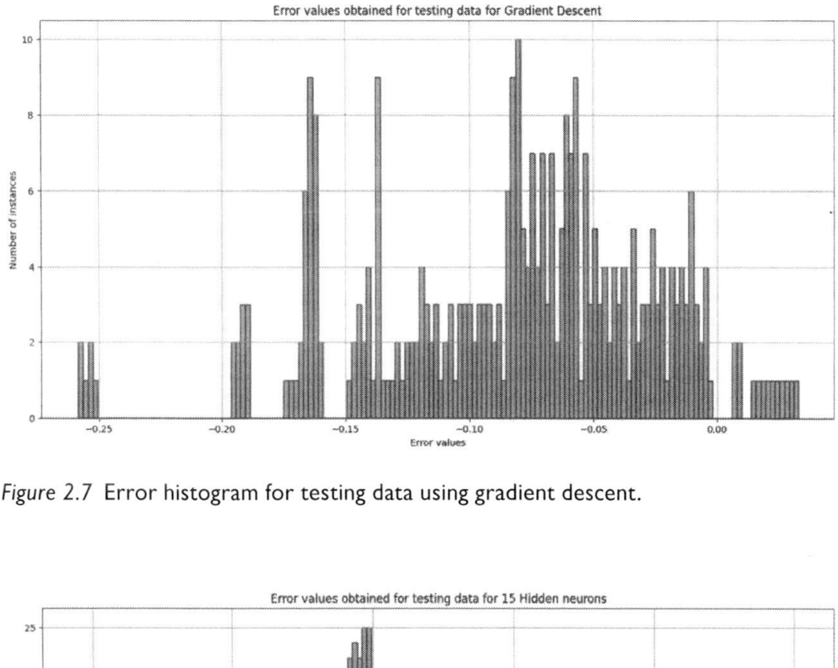

Figure 2.7 Error histogram for testing data using gradient descent.

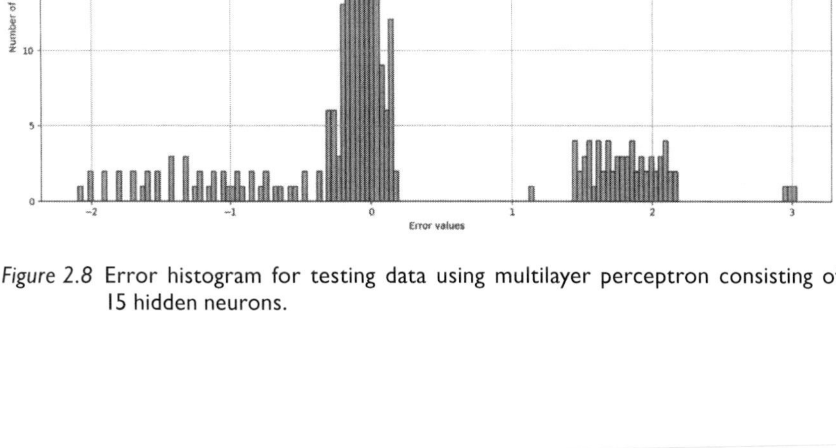

Figure 2.8 Error histogram for testing data using multilayer perceptron consisting of 15 hidden neurons.

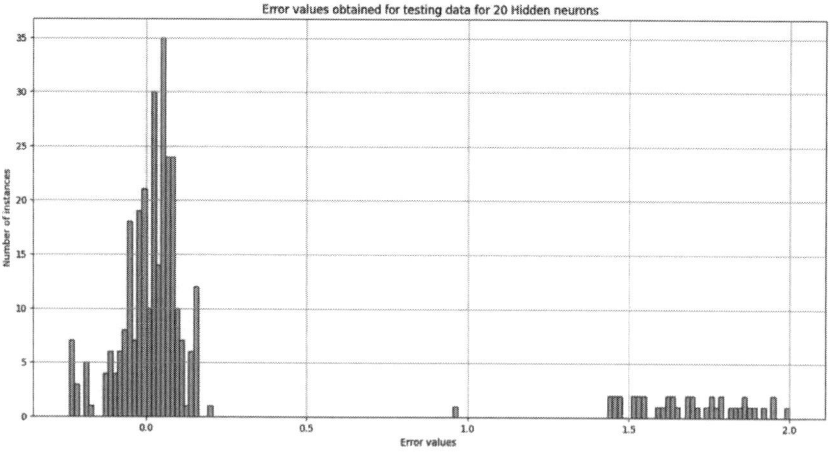

Figure 2.9 Error histogram for testing data using multilayer perceptron consisting of 20 hidden neurons.

The algorithm implemented below trains the data model to predict current at MPP on unseen data using the MLP algorithm. It plots the graph for the same and is also responsible for plotting a curve between the number of iterations and cost value per iteration.

We have adjusted the algorithm as per our requirement by introducing more hidden layers and using different activation functions for each hidden layer. We can also increase the number of neurons per hidden layer to make the algorithm perform well on unseen data and quickly discover more complex underlying hidden patterns. The study presented in this chapter shows better efficiency and faster convergence in the direction of minimum error while using a MLP with 20 neurons in the hidden layer compared to an MLP with 15 neurons in the hidden layer. The handling capability of neural networks on complex data samples with a suitable activation function is remarkable. After calculations, the mean absolute error for testing data using an MLP consisting of 20 neurons in the hidden layer is found to be 0.14075425, while the mean absolute error for testing data using an MLP composed of 15 neurons in the hidden layer is found to be 0.26291625. So, it is evident that an MLP with 20 neurons is more efficient than an MLP with 15 neurons.

The network's weights and biases are better adjusted by looping through multiple iterations with appropriate learning rates by passing over all of the training instances both forward and backward. It is incredibly effective for identifying a large number of intricate and invisible patterns in data sets, however the model is sophisticated due to the number of hidden layers and neurons. MPP is precisely determined under any fluctuation in atmospheric conditions using a neural network. Another advantage of adopting neural networks in PV MPP tracking is that they perform better dynamically than other approaches. The primary method of altering power is via modifying the duty cycle. The challenge of changing the duty cycle is exacerbated by uneven shading.

REFERENCES

1. Ruhi Sharmin, Sayeed Shafayet Chowdhury, Farihal Abedin, and Kazi Mujibur Rahman (2022), Implementation of MPPT technique of solar module with supervised machine learning, *Electrical Engineering and Systems Science*, pp. 1–11.
2. Theodore Amissah Ocran, Cao Junyi, Cao Binggang, and Sun Xinghua (2005), Artificial neural network maximum power point tracker for solar electric vehicle. *Tsinghua Science &* Technology, vol. 10, no. 2, p. 204–208.
3. Bidyadhar Subudhi, and Raseswari Pradhan (2013), A comparative study on maximum power point tracking techniques for photovoltaic power system. *IEEE Transactions on Sustainable Energy*, vol. 4, no. 1, pp. 89–98.
4. Adithya Ballaji, Ritesh Dash, Vivekanandan Subburaj, Kalvakurthi Jyotheeswara Reddy, Durgamadhab Swain, and Sarat Chandra Swain (2022), Design & development of MPPT using PSO with predefined search space based on fuzzy fokker planck solution. *IEEE Access*, vol. 10, pp. 80764–80783.
5. Mehmet Yilmaz and Muhammed Fatih Corapsiz (2022), Artificial neural network based MPPT algorithm with boost converter topology for stand-alone PV System. *Journal of Science and Technology*, vol. 15, no. 1, pp. 242–257.
6. Md Tahmid Hussain, Adil Sarwar, Mohd Tariq, Shabana Urooj, Amal BaQais, and Md. Alamgir Hossain (2023), An evaluation of ANN algorithm performance for MPPT energy harvesting in solar PV systems. *Sustainability*, vol. 15, no. 14, 11144.
7. Samer Gowid and Ahmed Massoud (2020), A robust experimental-based artificial neural network approach for photovoltaic maximum power point identification considering electrical, thermal and meteorological impact. *Alexandria Engineering Journal*, vol. 59, no. 5, pp. 3699–3707.
8. Žarko Zečević, and Maja Rolevski (2020), Neural network approach to MPPT control and irradiance estimation. *Applied Science*, vol. 10, no. 15, pp. 1–17.
9. Wang Jinpeng, Yao Qinxue, Zhang Bo, and Jeremy-Gillbanks And Zhao Xin (2022), Study on MPPT algorithm based on an efficient hybrid conjugate gradient method in a photovoltaic system. *IEEE Access*, pp. 4219–4227.
10. Karima Amara, Toufik Bakir, Ali Malek, Dalila Hocine, El-Bay Bourennane, Arezki Fekik, and Mustapha Zaoui (2019), An optimized steepest gradient based maximum power point tracking for PV control systems. *International Journal on Electrical Engineering and Informatics*, vol. 11, no. 4, pp. 662–683.
11. F. Bonanno, G. Capizzi, C. Napoli, G. Graditi, and G.M. Tina (2012), A radial basis function neural network based approach for the electrical characteris- tics estimation of a photovoltaic module. *Applied Energy*, vol. 97, pp. 956–961.

12. N. Ramesh Babu and Saravanan Subramani (2015), Performance analysis of boost Cuk converter in MPPT based PV system. *IEEE International Conf. on Circuit, Power and Computing Technologies*, pp. 1–6. doi:10.1109/ICCPCT.2015.7159425.
13. Shuran Jia, Daosheng Shi, Junran Peng, and Yang Fang (2015), Application of back-propagation neural network in multiple peak photovoltaic MPPT. *International Conference on Industrial Informatics - Computing Technology, Intelligent Technology, Industrial Information Integration*, pp. 231–234.
14. Bidyadhar Subudhi and Raseswari Pradhan (2013), A comparative study on max- imum power point tracking techniques for photovoltaic power systems.
15. Oladimeji Ibrahim, Nor Zaihar Yahaya, Nordin Saad, and Muhammad Wasif Umar (2015), Matlab/Simulink model of solar PV array with perturb and observe MPPT for maximising PV array efficiency.
16. Nayak, Jeeban Kumar, Arnab Ghosh, and Vikash Gurugubelli (2023), Maximum power point tracking using type-I fuzzy logic controller in PV system. In: *2023 IEEE 3rd International Conference on Sustainable Energy and Future Electric Transportation (SEFET)*, pp. 1–6. IEEE, New York.
17. S. Shyam and A. Ghosh (2023), Study on artificial neural network based MPPT algorithm in PV application. In: *2023 IEEE 3rd International Conference on Sustainable Energy and Future Electric Transportation (SEFET)*, pp. 1–6. IEEE, New York.
18. T. Anjum, J. K. Nayak, and A. Ghosh (2022), A comparative analysis between perturb observe and incremental conductance techniques for MPPT in PV system. In: *2022 IEEE 10th Power India International Conference (PIICON)*, pp. 1–6. IEEE, New York.
19. Sahu Rahul Kumar, and Arnab Ghosh (2022), Maximum power generation from solar panel by using P O MPPT. In: *2022 International Conference on Intelligent Controller and Computing for Smart Power (ICICCSP)*, pp. 1–4. IEEE, New York.
20. Roy, Aishworya, and Arnab Ghosh (2021), Current controlled grid-interfaced photo- voltaic boost inverter with particle swarm optimisation based MPPT algorithm. In: *2021 International Conference on Sustainable Energy and Future Electric Transportation (SEFET)*, pp. 1–5. IEEE, New York.
21. Kumar Nayak, Jeeban, Harika Thalla, and Arnab Ghosh. Efficient maximum power point tracking algorithms for photovoltaic systems with reduced number of sensors. *Process Integration and Optimization for Sustainability*, vol. 7, no. 1, pp. 191–213.
22. Aishworya Roy, Arnab Ghosh, Chiranjit Sain, Furkan Ahmad, and Luluwah Al- Fagih (2023), A comprehensive analysis of control strategies for enhancing regulation in standalone photovoltaic systems. *Energy Reports*, vol. 10, 4659–4678.
23. Joarder, Subhapriyo, Arnab Ghosh, Subrata Banerjee, Chiranjit Sain, and Furkan Ahmad (2024), Harmonic modelling and control of dual active bridge converter for DC microgrid applications. *Energy Reports*, 12, 52–74.
24. Gurugubelli, Vikash, Arnab Ghosh, and Anup Kumar Panda (2024), Improved Hopf oscillator-based VOC method for fast synchronization of parallel inverters in standalone microgrid. *IEEE Journal of Emerging and Selected Topics in Power Electronics*, vol. 12, no. 3, 3018–3025.

25. Reddi, Charan, and Arnab Ghosh (2024). Bidirectional DC converter using Li-Ion battery for EV applications. In: *2024 IEEE 13th International Conference on Communication Systems and Network Technologies (CSNT)*, pp. 1270–1275. IEEE, New York.
26. Tiwari, Harshita, Arnab Ghosh, Chiranjit Sain, Furkan Ahmad, and Luluwah Al-Fagih (2024), Modified direct torque control algorithm for regeneration capability of IM driven electric vehicle by using hybrid energy storage system. *Renewable Energy Focus*, 48, 100534.
27. Barker, Teja, Arnab Ghosh, Chiranjit Sain, Furkan Ahmad, and Luluwah Al-Fagih (2024), Efficient ANFIS-driven power extraction and control strategies for PV-bess integrated electric vehicle charging station. *Renewable Energy Focus*, 48, 100523.
28. Shubham Sinha and Arnab Ghosh (2024), Switching transients reduction for battery charging controller during mode selection. In: *2024 IEEE International Students' Conference on Electrical, Electronics and Computer Science (SCEECS)*, pp. 1–6. IEEE, New York.
29. Shubham Sinha and Arnab Ghosh (2024), TPUPS topology implemented with model predictive control. In: *2024 Third International Conference on Power, Control and Computing Technologies (ICPC2T)*, pp. 535–540. IEEE, New York.
30. Arnab Ghosh, Annesha Sahu, and Puja Barua (2024), Particle swarm optimization-based MPPT for PV applications and its comparison with incremental conductance method. In: *Optimization Techniques for Hybrid Power Systems: Renewable Energy, Electric Vehicles, and Smart Grid*, pp. 308–324. IGI Global, Hershey.

Chapter 3

An alternative approach to the modeling of harmonic impedance of a salient pole synchronous generator

Adhir Baran Chattopadhyay, Sunanda Hazra, Rishabh Dev Shukla, and Sudeshna Nath

ABBREVIATION

V_{ah}, V_{bh}, V_{ch} – harmonic voltages of harmonic order, "h" in phases a, b and c, respectively

Re – real component of any complex number

β – initial phase angle of per phase voltage injection

t – time (s)

The instantaneous position of the rotor is,

$\theta = \omega t + (\pi/2) + \delta$, where

δ=load angle (power angle) and ω=frequency in electrical rad/s

V_{dh}, V_{qh}, V_{0h} – Harmonic voltages of harmonic order, "h" – d-axis, q-axis and zero-sequence components, respectively

i_{dh}, i_{qh} – harmonic currents (axis components)

C''_d , C''_q – fictitious d-axis and q-axis subtransient capacitance parameters, respectively

G – the conductance parameter

$X''_d = \omega L''_d$ and $X''_q = \omega L''_q$ – d-axis and q-axis subtransient reactances, respectively

L''_d and L''_q – Equivalent subtransient inductances along d-axis and q-axis, respectively.

INTRODUCTION

It is well known that the behavior of a power system regarding the involvement of harmonic components basically depends upon the harmonic impedance modeling of the salient pole synchronous generator. It is also well known that during the steady-state operation of a salient pole generator, the fundamental frequency negative sequence currents carried by the armature windings lead to the development of voltages of fundamental frequency as

 DOI: 10.1201/9781003628279-3

well of third harmonic frequency [1–4]. Such behavior of the machine is because a negative sequence current with fundamental frequency "ω" in the armature corresponds to sustained oscillation of positive sequence d-axis and q-axis voltages in the rotor with "2ω" frequency superimposed on the fact that the mechanism of transformation from axis quantities to phase quantities represents another sustained oscillation with "ω" frequency. Similarly, negative sequence applied voltages give rise to currents of fundamental frequency and third harmonics frequency.

Generally, power system analysis is performed using electromagnetic transient programs, such as Electromagnetic Transient Program (EMTP) [5] and Electromagnetic Transients including DC (EMTDC) [6], with the synchronous generator using Park's transformation [7–9]. Generally, when the voltages and currents of a salient pole synchronous generator are not directly specified, but rather specified through power system data, both voltages and current must be expressed by an infinite series of harmonics. However, the behavior of the salient pole generator under transient conditions becomes difficult when the simultaneous influences of the terms $e^{\alpha t}$ and sin (ωt) are to be considered, where α represents the reciprocal of the time constant of DC component, and "ω" represents the frequency (rad/s) of the term corresponding to sustained oscillation. Hence, there is a necessity that the time–domain models of the machine should be allowed to reach a steady state such that analyzing the harmonic behavior of the machine becomes feasible. From this viewpoint, the electromagnetic transient program in the references [5,6] appears to be an inefficient process because to reach a steady state, such programs demand a long time gap.

A generator model formulated in phase components has been described by Chattopadhyay et al. [9] that takes rotor saliency into account in the harmonic domain. However, this model does not include the effect of rotor angle.

There is a companion paper [10] on the modeling of a salient pole synchronous generator for harmonic analysis. The authors of the paper [10–12] derived the expressions for the harmonic impedances of the generator due to the negative sequence harmonic currents injected into the machine in a nice manner. In other words, they have derived the equation by the current reference method [13,14]. But from the viewpoint of harmonic impedances, it is known that due to the negative sequence harmonic voltages injected into the machine, harmonic impedances can develop. This viewpoint has not been taken care by the authors of the said paper [6], and the work in this particular direction forms the subject matter of this chapter. However, Adkins and Harley [1] have dealt with the mathematical treatment of the negative sequence reactance of the salient pole synchronous machine using both the "voltage reference" method and the "current reference" method. They have obtained different expressions for negative sequence reactance using the above-said methods, and this appears logically correct, but to reach the convenient form of the expression, they have considered one assumption, which is based on the theory of large slip. The authors of this chapter do not use the theory of large slip to derive the expressions for the harmonic impedances, rather they approach from the viewpoint of the theory of symmetrical components and Park's transformation. As a consequence, the so-called new parameters, such as subtransient capacitance (C_d'' and C_q'' along *d*-axis and *q*-axis, respectively), have been introduced in the proposed mathematical formulation.

The concept of subtransient capacitances may appear fictitious at first glance because, generally, terminologies like subtransient/transient reactances (inductances) are mostly popular in association with electrical machines as these machines are built based on magnetic field energy storage systems. It may be recalled in this context that the historically older machine, Van-De-Graff generator was practically built using an electric field as the coupling medium. Variable capacitor also works as the energy storing medium to generate a certain amount of electrical power [15]. The existence of "transient capacitance" or "subtransient capacitance" in electromechanical devices or electrical machines is very rare. However, Kim and Shin [16] deal with the ferroelectric capacitor used in complementary metal-oxide-semiconductor (CMOS) technology, and it shows a transient negative capacitance effect.

There is another research work by Chang et al. [17] that investigated the origin of transient negative capacitance in a ferroelectric capacitor. It probes the foundation of ferroelectricity. In this paper, a resistor-ferroelectric capacitor (R-Fe C) network through a series of coupled equations based on Kirchhoff's law, electrostatics and Landau theory has been studied.

The significance of the present problem formulation technique is presented in a separate section for a better understanding of this research problem as follows.

THE SIGNIFICANCE OF THE FORMULATION TECHNIQUE

In the last paragraph of the Introduction section, it has been mentioned that the theory of large slip was used by Adkins and Harley [1]. The main difficulty lies in the said assumption that the machine, which is originally a synchronous machine, has been assumed to be an induction machine with a uniform air-gap. Hence, such a proposition contradicts the magnetic asymmetry of a synchronous machine (generator). Once a machine is considered to have a non-uniform air-gap, then the theory of slip to apply while formulating the research problem may not give the correct picture of the armature reaction phenomenon of a machine based on a multi-excited magnetic field system. However, it is also conceptually true that to apply *d*–*q* resolution technique to an induction machine (which is basically a uniform air-gap machine) under sudden disturbance conditions is not fundamentally wrong because even though the physical structure of an induction machine maintains uniform air-gap throughout the rotor periphery, under subtransient or transient conditions, effectively the air-gap permeances along two orthogonal axes of the machine become unequal to maintain "Constant flux linkage theorem" (also due to the additional fact of different values of "the ratio of inductance to resistance" of large number of small eddy current loops formed on the two orthogonal axes [14]). Such differences in the air gap lengths along those axes lead to the development of *d*–*q* resolution approach. Here lies the novelty of the philosophy used in the proposed research work for mathematical modeling of the phenomenon of development of harmonic impedance in a three-phase salient pole synchronous generator (machine) when a negative sequence voltage set is injected.

The crux of mathematical modeling of the above-said impedance phenomenon lies in the fact that based on certain voltage if we want to find out any current under the presence of both energy storing element and energy dissipating element, then we will have no other fundamental mechanism than to adopt an R-C (resistance and capacitance in parallel combination) circuit. As a matter of fact, this fundamental circuit theory concept becomes the foundation of the problem formulation. Once this philosophy is made clear, then automatically, one question that arises in the mind is whether it will correspond to the steady-state or transient/subtransient phenomena. The answer to such a question is dictated by the fact that the negative sequence reactance of a salient pole synchronous generator is conceptually related to subtransient reactances along the direct axis (*d*-axis) and quadrature axis (*q*-axis). So, the formulation based on the R-C circuit approach must include subtransient capacitance parameters. Such inclusions have been executed during the present formulation.

Another important aspect of the proposed mathematical formulation is how to realize or simulate a capacitance in an electrical machine, the function of which is based on a magnetic field energy storage system, not on the electric field storage system. The way to come out of such a dilemma is shown by the use of the "electrical resonance concept" which can easily represent a capacitance parameter by an equivalent inductance parameter involving the important circuit variable which is "supply frequency" or "induced frequency" used as per the actual phenomenon. Such philosophy has been exactly used in the problem formulation presented in the next section.

The finer tuning of the problem formulation is embedded in the conversion of frequency value when the electrical resonance principle is applied. An electrical speed "ω" of any positive sequence field must correspond to the value, "2ω" of the negative sequence field. This is the basis of the said finer tuning.

The "reference frame theory" (which is a major concept of generalized theory of electrical machines) has been used in the present problem formulation to avoid the "theory of large slip" as pointed out in the last paragraph of the Introduction section. Basically, the axis harmonic currents (both on the d-axis and the q-axis) have been formulated after obtaining the expressions of axis components of harmonic voltages based on the application of Park's transformation matrix (which is one of the major reference frames used so far). Once the voltage-dependent axis harmonic currents are formulated involving the resistance parameters and the fictitious capacitance parameters, again the inverse Park's transformation matrix is applied to bring the axis currents back into the phase model of the synchronous machine, as a function of the phase voltages. Then, automatically, the expressions for harmonic admittances or reactances come out.

The above-said paragraphs constitute the conceptual mechanism behind the whole problem formulation presented in the next section.

PROBLEM FORMULATION

Generator negative sequence harmonic admittances are derived by injecting negative sequence harmonic voltages into generator Park's transformation equations and taking ratios of the resultant currents and terminal voltages at each harmonic order "h." From hereafter, this method is designated as the "voltage reference" method. For a harmonic order "h," negative sequence voltage injections are

$$\begin{bmatrix} V_{ah} \\ V_{bh} \\ V_{ch} \end{bmatrix} = \mathrm{Re}\begin{bmatrix} Ve^{j(h\omega t-\beta)} \\ Ve^{j\left(h\omega t+\frac{2\pi}{3}-\beta\right)} \\ Ve^{j\left(h\omega t-\frac{2\pi}{3}-\beta\right)} \end{bmatrix} \tag{3.1}$$

where β=initial phase angle of per phase voltage injection.

The field winding is short-circuited, since it is not excited at harmonic frequency, i.e.,

$$V_{\mathrm{fh}} = 0 \tag{3.2}$$

The field rotates at synchronous speed with d-axis leading phase "a" positive sequence voltage as shown in Figure 3.1.

The instantaneous position of the rotor is,

$$\theta = \omega t + (\pi / 2) + \delta \tag{3.3}$$

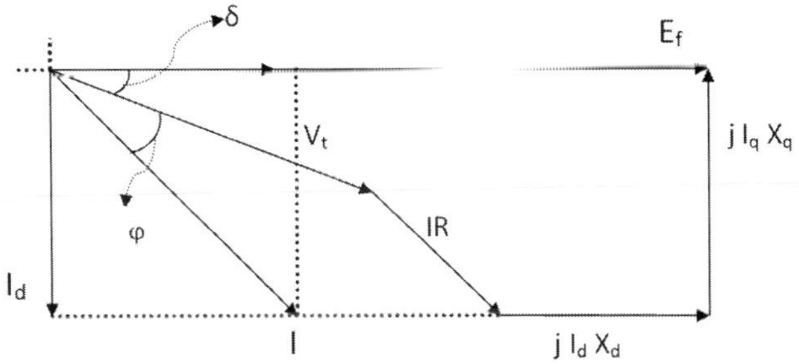

Figure 3.1 Phasor diagram of a salient pole synchronous generator.

Park's transformation from phase model (a, b, c) to axis model (d, q) can be expressed as,

$$\begin{bmatrix} V_{dh} \\ V_{qh} \\ V_{oh} \end{bmatrix} = \frac{2}{3} \begin{bmatrix} \cos\theta & \cos(\theta - 2\pi/3) & \cos(\theta - 2\pi/3) \\ -\sin\theta & -\sin(\theta - 2\pi/3) & -\sin(\theta + 2\pi/3) \\ 1/2 & 1/2 & 1/2 \end{bmatrix} \begin{bmatrix} V_{ah} \\ V_{bh} \\ V_{ch} \end{bmatrix} \tag{3.4}$$

Substituting equation (3.1) in equation (3.4) yields,

$$V_{dh} = \frac{2}{3}\left[V_{ah}\cos\theta + V_{bh}\cos(\theta - 2\pi/3) + V_{ch}\cos(\theta + 2\pi/3\right] \tag{3.5}$$

$$\begin{bmatrix} V_{dh} \\ V_{qh} \\ V_{oh} \end{bmatrix} = V \begin{bmatrix} \cos\left[(h+1)\omega t + (\pi/2) + \delta - \beta\right] \\ \sin\left[(h+1)\omega t + (\pi/2) + \delta - \beta\right] \\ 0 \end{bmatrix} \begin{bmatrix} V_{ah} \\ V_{bh} \\ V_{ch} \end{bmatrix} \tag{3.6}$$

As the objective is to derive the expression for harmonic impedance or admittance, at this stage, it becomes necessary to formulate the expression for harmonic currents along d-axis and q-axis. Hence, for d-axis, it yields,

$$i_{dh} = GV_{dh} + C_d''\frac{dV_{dh}}{dt} \tag{3.7}$$

where "G" represents the conductance parameter and C_d'' represents fictitious d-axis subtransient capacitance parameter. C_d'' may be expressed by using the condition of resonance, as given by,

$$\omega C_d'' = \frac{1}{2\omega L_d''} \tag{3.8}$$

where L_d'' represents the d-axis subtransient armature inductance. Equation (3.8) indicates that a negative sequence field with frequency "ω" (in electrical rad/s) corresponds to a positive sequence field of frequency "2ω" (electrical rad/s) in a rotor reference frame. From equations (3.7) and (3.6), i_{dh} can be expressed as

$$i_{dh} = GV\cos\left(\theta+(h)\omega t-\beta\right) - \omega C_d''\ V\left((h)+1\right)\left(\sin\left(\theta+(h)\omega t-\beta\right)\right) \tag{3.9}$$

where

$$\theta = \omega t + (\pi/2) + \delta \tag{3.10}$$

Similarly,

$$i_{qh} = GV_{qh} + C_q''\frac{dV_{qh}}{dt} \tag{3.11}$$

where C_q'' is fictitious q-axis subtransient capacitance parameter, given by

$$\omega C_q'' = \frac{1}{2\omega L_q''} \tag{3.12}$$

where L_q'' represents the q-axis subtransient armature inductance.

$$i_{qh} = -GV\sin\left(\theta+(h)\omega t-\beta\right) - \omega C_q''\ V\left((h)+1\right)\left(\cos\left(\theta+(h)\omega t-\beta\right)\right) \tag{3.13}$$

Equations (3.8)–(3.12) involve the appearance of the capacitance parameters C_d'' and C_q''. From the viewpoint of the operation of the synchronous machine, such parameters do not carry any physical significance. However, equations (3.7)–(3.11) represent current balance equations, which are more relevant to an electrostatic field energy system. As inherently the electrostatic field involves the capacitance parameters, in the present analysis also, the fictitious parameters C_d'' and C_q'' have been forcibly introduced. However, the analysis of synchronous machines in various operating conditions is generally carried out using different types of inductance parameters, and it is also quite logical because the coupling field of a synchronous machine is a magnetic field system.

It is well known that the subtransient reactances in synchronous machines are developed because the constant flux linkage theorem is maintained by the direct-axis or quadrature-axis damper windings. If, in the dual mode of configuration, a similar synchronous machine is assumed based on an electrostatic field as a coupling medium, C_d'' and C_q'' can be realized physically.

At this stage, the authors reveal that the concept of subtransient capacitance can be physically explained in detail as shown in Figure 3.2a and b. The analogy can be considered similar to that of a variable capacitance system [11]. Figure 3.2a shows a variable capacitance system. The switch is closed initially. Hence, an equal and opposite charge develops across the plates of the capacitor. In Figure 3.2b, the source is open circuited. This leads to the development of an intermediate charge, equal and

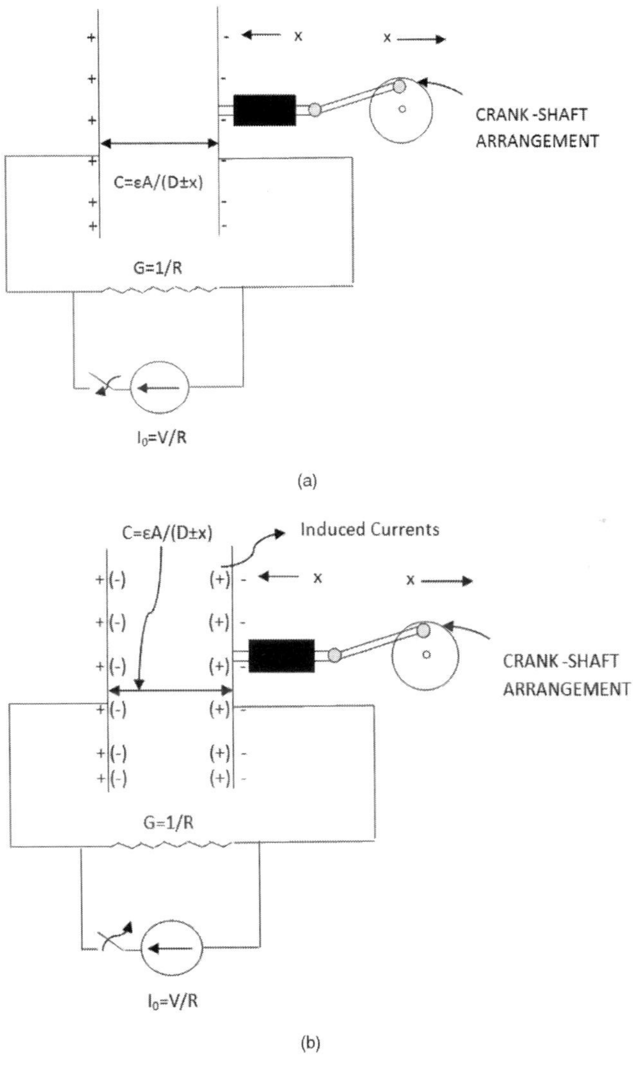

Figure 3.2 (a) Realization circuit of C_d'' and C_q'' before perturbation (switch closed). (b) Realization circuit of C_d'' and C_q'' during perturbation (switch open).

opposite in magnitude, on the inner and outer surfaces of both the capacitor plates. Hence, the single plate can now be visualized as a capacitor itself. This is subtransient capacitance during perturbation. Also, a change in charge magnitude is equal to a change in flux. For the system in consideration, the change in charge magnitude is zero. This corresponds to no change in flux. Hence, constant flux linkage theorem is established. Hence, due to necessity, while developing current balance equations (3.7) and (3.11), the forcible introduction of the fictitious parameters C_d'' and C_q'' is a must, but such equation cannot be carried on for further analysis of the machine. To tackle such a critical situation, the condition of resonance has been used in equations (3.8) and (3.12) by adjusting the frequency value, such that the equivalent inductance parameters come into picture for carrying out further analysis of the synchronous machine.

Now, using axis (d, q) model to phase (a, b, c) model transformation, it yields

$$\begin{aligned}
i_{ah} &= i_{dh}\cos\theta - i_{qh}\sin\theta \\
&= GV\ \cos\theta\cos(\theta + h\omega t - \beta) - \omega C_d''\ V(h+1)\cos\theta\sin(\theta + h\omega t - \beta) \\
&\quad + GV\sin\theta\sin(\theta + (h)\omega t - \beta) + \omega C_q'' V(h+1)\sin\theta\cos(\theta + (h)\omega t - \beta) \\
&= GV[\cos\theta\cos(\theta + (h)\omega t - \beta) + \sin\theta\sin(\theta + (h)\omega t - \beta)] \\
&\quad + \omega C_q'' V\frac{(h+1)}{2}\ [2\sin\theta\cos(\theta + (h)\omega t - \beta)] \\
&\quad - \omega C_d'' V\frac{(h+1)}{2}[2\cos\theta\sin(\theta + (h)\omega t - \beta)]
\end{aligned} \tag{3.14}$$

By substituting equations (3.8) and (3.12) in equation (3.14), we obtain

$$\begin{aligned}
i_{a(h)} &= GV\cos\big((h)\omega t - \beta\big) + V\ \frac{\big((h)+1\big)}{2}\left(\frac{X_d'' - X_q''}{2\ X_d''\ X_q''}\right) \\
&\quad \times\sin\Big(\big((h)+2\big)\omega t + \pi + 2\delta - \beta\Big) \\
&\quad - V\ \frac{\big((h)+1\big)}{2}\sin\big((h)\omega t - \beta\big)\left(\frac{X_d'' + X_q''}{2\ X_d''\ X_q''}\right)
\end{aligned} \tag{3.15}$$

where

$$X_d'' = \omega L_d'' \tag{3.16a}$$

and

$$X_q'' = \omega L_q'' \tag{3.16b}$$

Equation (3.15) can be expressed in a different form given by

$$\begin{aligned} i_{ah} &= \text{Re}\left[G + j\,\frac{(h+1)}{2}\left(\frac{X_d'' + X_q''}{2\,X_d''\,X_q''}\right)Ve^{j(h\omega t-\beta)}\right] \\ &+ \text{Re}\left[V\,\frac{(h+1)}{2}\,je^{j((h+2)t+\pi+2\delta-)}\left(\frac{X_q'' - X_d''}{2\,X_d''\,X_q''}\right)\right] \end{aligned} \tag{3.17}$$

$$\text{Let, } \theta_1 = \pi + 2\delta - \beta \tag{3.18}$$

such that,

$$e^{j\theta_1} = e^{j\left(\frac{\pi}{2}+2\delta-\right)}e^{j\frac{\pi}{2}} \tag{3.19}$$

Hence, substituting equation (3.19) in equation (3.17) yields

$$\begin{aligned} i_{a(h)} &= \text{Re}\left[G + j\,\frac{((h)+1)}{2}\left(\frac{X_d'' + X_q''}{2\,X_d''\,X_q''}\right)Ve^{j((h)\omega t-\beta)}\right] \\ &- \text{Re}\left[V\,\frac{((h)+1)}{2}\,je^{j\left(((h)+2)t+\frac{\pi}{2}+2\delta-\beta\right)}\left(\frac{X_q'' - X_d''}{2\,X_d''\,X_q''}\right)\right] \end{aligned} \tag{3.20}$$

The ratios of currents and voltages in the same harmonic order are the harmonic self-admittances,

$$Y_{-h,-(h)} = G + j\,\frac{(h+1)}{2}\left(\frac{X_d'' + X_q''}{2\,X_d''\,X_q''}\right) \tag{3.21}$$

If h is substituted equal to 1 in equation (21), then the expression for fundamental frequency negative sequence susceptance (B_n) comes out to be,

$$B_n = \left(\frac{X_d'' + X_q''}{2\,X_d''\,X_q''}\right) \tag{3.22}$$

which exactly matches with the formulation as reported by Adkins and Harley [1]. The generators cross couplings from negative sequence harmonic voltages h to positive sequence harmonic currents $h+2$ are

$$Y_{+(h+2),-h} = \frac{(h+1)}{2}\left(\frac{X_q'' - X_d''}{2\ X_d''\ X_q''}\right) e^{j\left(2t+\frac{\pi}{2}+2\delta\right)} \tag{3.23}$$

An illustrative schematic representation of the total concept may be visualized closer to the method. It is presented in the subsequent section.

An illustrative schematic representation of the total concept

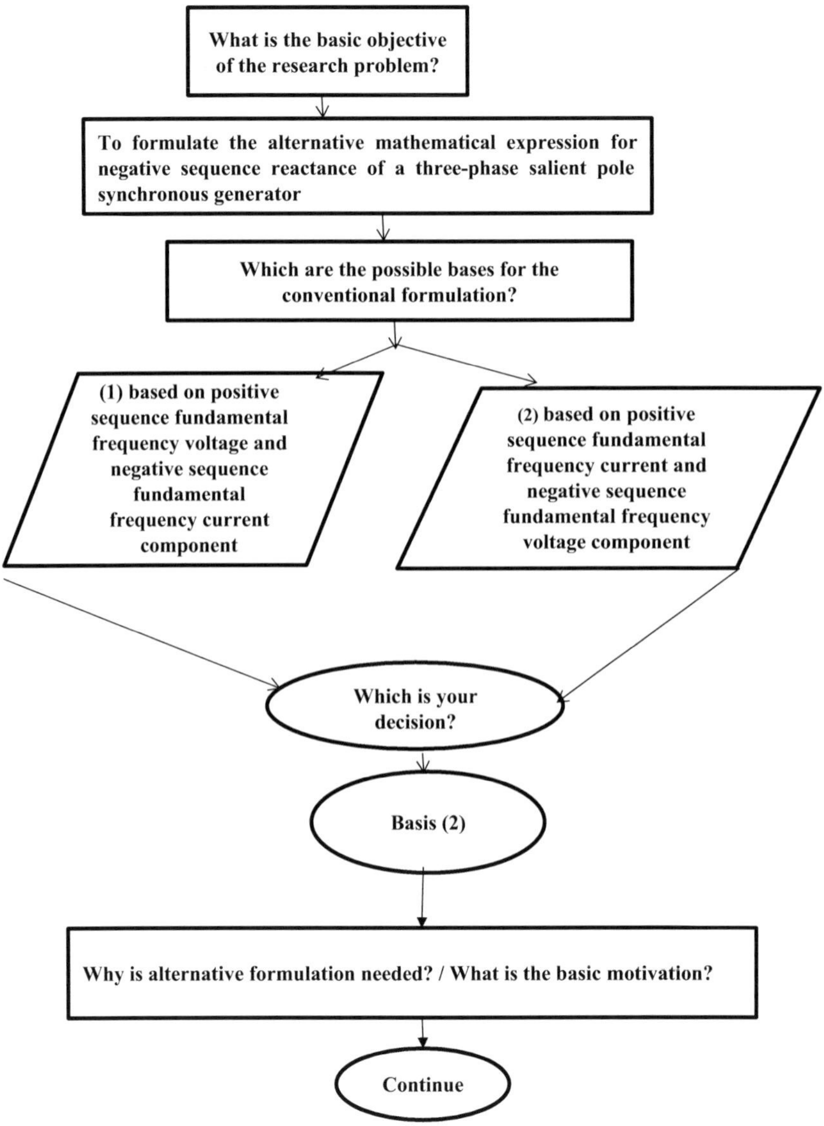

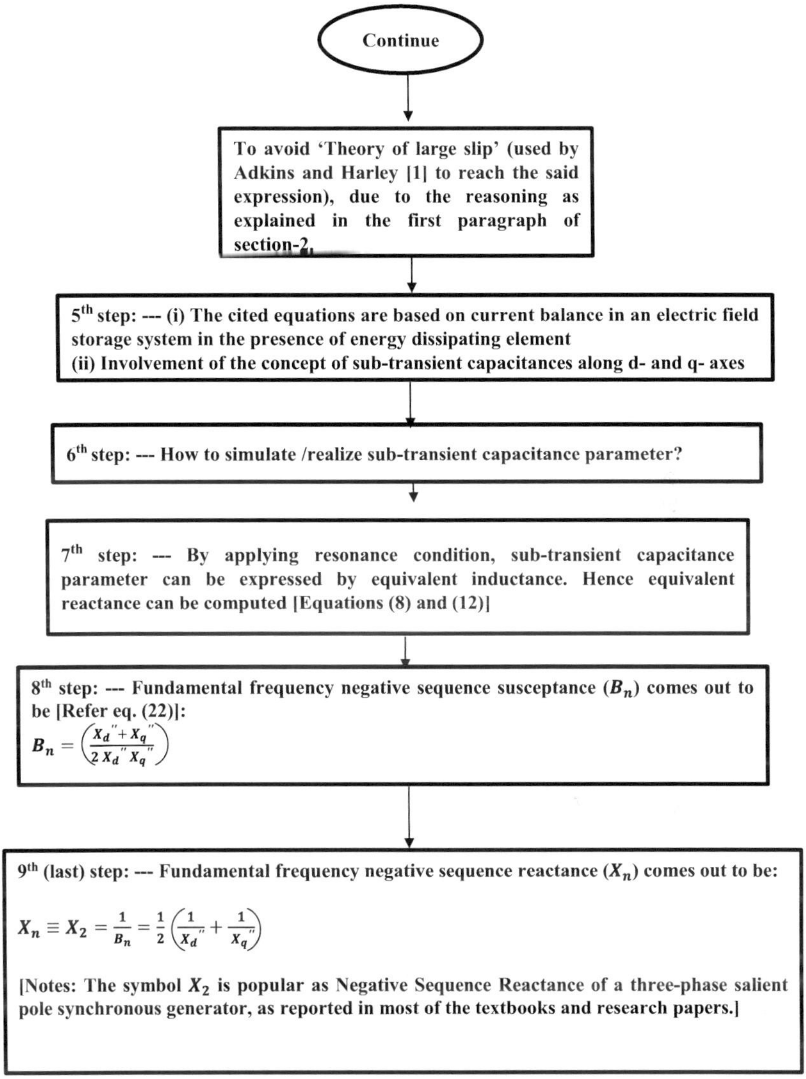

RESULTS AND DISCUSSIONS

The analysis based on the negative sequence fundamental frequency voltage component leads to the formulation for the negative sequence susceptance, which has been expressed in equation (3.22). The reciprocal of this susceptance must be a reactance parameter, and this can be expressed as

$$X_{2(2)} = \frac{2\left(X_d''\right)\left(X_q''\right)}{\left(X_d''\right)+\left(X_q''\right)} \tag{3.24}$$

For verification of the correctness of the formulation for such reactance, the computed values using equation (3.24) must be compared with a standard expression for the same reactance. This standard expression reported in [1] is based on positive sequence fundamental frequency voltage and negative sequence fundamental frequency current component. This reactance may be named as $X_{2(1)}$, and it has been formulated as

$$X_{2(1)} = \frac{\left(X_d''\right)+\left(X_q''\right)}{2} \tag{3.25}$$

As generally, third harmonic components of the three-phase currents of a generator (or, power system) are co-phasal in nature; they may be compared with the zero-sequence current components of an unbalanced system. Furthermore, these third harmonic current components of positive sequence, interacting with negative sequence fundamental frequency voltage, lead to the development of cross-coupling reactance which must be computed for the successful design of a synchronous generator. Based on equation (3.23), such cross-coupling reactance can be expressed as,

$$X_{3,-1} = \frac{2\left(X_d''\right)\left(X_q''\right)}{\left(X_q''\right)-\left(X_d''\right)} \tag{3.26}$$

Based on equations (3.24)–(3.26), the numerical computation work has been completed, and these are presented in Table 3.1. The standard design values of the basic parameters like X_d'', X_q'', $\omega C_d''$ and $\omega C_q''$ have been taken from different standard literature as shown in Table 3.1, and these parameter values are used to compute the reactances expressed in equations (3.24)–(3.26).

The existence of the "transient capacitance" or "subtransient capacitance" in electromechanical devices or electrical machines is very rare. However, with reference to Table 3.1, the reference literature [12] deals with the ferroelectric capacitor used in CMOS technology, and it shows a transient negative capacitance effect.

With reference to Table 3.1, the negative sign associated with the value of the generator cross-coupling reactance $X_{(3,-1)}$ indicates the fact which is $\left(X_q''\right)-\left(X_d''\right)<0$. Again, from the viewpoint of machine design, such inequality becomes valid only when the length of leakage flux paths of field winding and direct-axis damper winding are kept in the low-value regions.

Table 3.1 Results of computation work for the reactance values expressed in equations (3.24)–(3.26)

Serial no	X''_d	X''_q	$\omega C''_d$	$\omega C''_q$	$X_{2(1)} = \frac{(X''_d) + (X''_q)}{2}$	$X_{2(2)} = \frac{2(X''_d)(X''_q)}{(X''_d) + (X''_q)}$	$X_{3,-1} = \frac{2(X''_d)(X''_q)}{(X''_q) - (X''_d)}$
1 [11]	0.204	0.237	2.45	2.1097	0.220	0.219	2.930
2 [12]	0.1831	0.1676	2.7307	2.9833	0.1750	0.1754	–3.959
3 [13]	0.135	0.2	3.7037	2.50	0.1675	0.1612	0.830
4 [13]	0.398	0.427	1.256	1.170	0.412	0.4125	11.720

CONCLUSION

This chapter highlights the following concluding points:

a. The alternative approach of modeling harmonic impedance presented in this chapter may act as a catalyst to improve the post-graduate teaching methodology.
b. The concept of a fictitious capacitance parameter and the way of converting this parameter into an equivalent inductance parameter by using the concept of resonance may be useful as a teaching module in other areas of electrical engineering.
c. The introduction of constant flux linkage theorem pertaining to a dielectric medium can open up new horizon for analysis.
d. The sudden short circuit phenomenon in an energy system based on a magnetic field resembles the sudden open circuit phenomenon in an energy system based on an electrostatic field. Hence, the concept of duality can be deeply analyzed in transient situations as it is a perturbation problem.
e. It is well known that the analysis of a synchronous generator, due to a sudden short circuit imposed at the armature winding terminals, is usually carried out either by small perturbation theory or by superposition theorem. It will be a matter of interest to see if the resulting model is suddenly open circuited.

REFERENCES

1. B. Adkins and R.C. Harley, *The General Theory of Alternating Current Machines-Application to Practical Problem*, pp. 143–145, Chapman and Hill, London, 1975.
2. C. Concordia, *Synchronous Machine: Theory and Performance*, Wiley, New York, 1951.
3. E.W. Kimbark, *Power System Stability: Synchronous Machines*, Dover Publication, New York, 1968.
4. W.C. Duesterhoeft, "The negative sequence reactance of an ideal synchronous machine", *AIEE Transactions*, 68(1), 1949, pp. 510–514.
5. H.W. Dommel, "Digital computer solution of electromagnetic transients in single and multiphase network", *IEEE Transaction on Power Apparatus and Systems*, 88(4), 1969, pp. 388–399.
6. D.A. Wooford, A.M. Gole and R.W.Menzies, "Digital simulation of D.C. links and A.C. machines", *IEEE Transaction on Power Apparatus and Systems*, 102(6), 1983, pp. 1616–1623.
7. R.H. Park, "Two reaction theory of synchronous machines-Part 1: generalized method of analysis", *AIEE Transactions*, 48, 1929, pp. 716–730.
8. C.V. Jones, *The Unified Theory of Electrical Machine*, Butterworths, London, 1967.

9. S. Chattopadhyay, M. Mitra and S. Sengupta, "Harmonic analysis in a three phase system using park transformation technique modelling, measurement and control", *American Society of Mechanical Engineers*, 80(3), 2007, pp. 42–57.
10. A. Medina and J. Arrilaga, "Analysis of transformer-generator interaction in harmonic domain", *IEE Proceedings. Generation, Transmission, and Distribution*, 141(1), 1994, pp.38–46.
11. A.B. Chattopadhyay, S. Thomas and R. Chatterjee, "Analysis of steady state stability of a CSI fed synchronous motor drive system with damper windings included", *Trends in Applied Sciences Research*, 6(9), 2011, pp. 992–1005.
12. https://www.skm.com/Synchronous_Generator_Input_Data.html
13. K.R. Padiyar, *Power System Dynamics and Control*, 2nd edition, pp. 80–81, B.S. Publications, India, 2012.
14. I.M. Canay, "Extended synchronous-machine model for the calculation of transient processes and stability", *Electric Machines & Power Systems*, 1(2), 1977, pp. 137–150.
15. J.Y. Dea, "Variable capacitor based mechanical to electrical generator", https://patents.google.com/patent/US9184675B1/en; 2011.
16. T. Kim and C. Shin, "Effects of interface trap on transient negative capacitance effect: phase field model", *Electronics*, 20209, 2020, p. 2141. doi:10.3390/electronics9122141; www.mdpi.com/journal/electronics
17. S.-C. Chang, U.E. Avci, D.E. Nikonov, S. Manipatruni, I.A. Young, "Physical origin of transient negative capacitance in a ferroelectric capacitor", *Physical Review Applied*, 9, 2018, p. 014010.

Chapter 4

Neuro-fuzzy–based optimization techniques

Asit Kumar Nayek, Jibendu Kumar Mantri, Arpan Adhikary, and Sunanda Hazra

INTRODUCTION

Energy usage pattern is a developmental indicator of a nation. Per capita energy requirement is high in developed countries and considerably low for underdeveloped countries. Optimized use of energy is a primary concern to resource managers since fossil fuels are limited to this planet and generating electricity from these resources creates the problem of global warming and CO_2 production [1]. Our motive is to maximize energy production from fossil fuels and renewable energy sources and to minimize power transmission loss.

Not only in the energy segment, almost all engineering problems are associated with optimization (either maximizing or minimizing) issues. With the advancement of computer and information technology, the job of data analysis has become easier. In electro-mechanical engineering problems, the application of fuzzy inference system plays a significant role. In dynamic factories, many constraints influence quality maintenance. These parametric settings are often considered uncertain, and to make a good prediction out of these parameters, neuro-fuzzy convex optimization technique is adopted [2,3].

In this chapter, we try to explore the recent techniques of optimization using soft computing techniques. Relevant recent papers are collected and their principles are summarized in the literature review subsection. Thereafter, classical adaptive neuro-fuzzy inference system (ANFIS), classical particle swarm optimization (PSO), and its recent variant, craziness-based particle swarm optimization (CRPSO) have been discussed. A relevant topic, recursive least square (RLS) algorithm has been discussed. The characteristics of neuro-fuzzy convex optimization have been discussed. We conclude this chapter by comparing the performance and applicability of these algorithms.

 DOI: 10.1201/9781003628279-4

LITERATURE REVIEW

In machine learning, extracting features or finding important and deciding attributes of objects is a challenging task. Finding correlation among different features is considerably important when the data analysis task is initiated. Ritam Guha et al. (2020) [4] introduced a version of nature-inspired algorithm, clustering-based population in binary gravitational search (CPBGS), that was applied to clustering problems in unsupervised mode. However, this CPBGS algorithm was not tested against feature selection problems. Fuzzy inference system (FIS) was used to design fuzzy controllers to optimize home-based photovoltaic microgrids. Derived fuzzy rules were optimized using ANFIS. In the last two decades, renewable energy generation has increased many folds, and now, we are trying to integrate conventional fossil fuel energy with renewable energy sources with the help of microgrid arrays. As renewable energy sources have some irregularities with respect to frequency, amplitude, etc., the use of simple proportional integral derivative (PID) controllers is not a feasible solution. To make a useful PID controller in such a scenario, fine-tuned PID controllers are used. These controllers are tuned by neuro-fuzzy-system, and the previously captured data has been trained using artificial neural network (ANN). In this context, fuzzy self-tuning single neuron PID was proposed in [4]. Maximum power tracking (MPT) is also a solution to heterogeneous grids. And parameters associated with the MPT could be optimized using ANFIS. PSO was used to train the ANFIS, and the least square algorithm was used to update the parametric values [5]. In optimization problems, hyper-parameters need to be optimized, and for that, bio-inspired algorithms are introduced. This kind of associative computing is commonly known as soft computing. In certain cases, these hyper-parameters don't follow crisp logic. Hence, to fit into the soft computing model, they are modeled in neuro-fuzzy hybrid systems. Discrete particle swarm optimization (DPSO), a variant of PSO, was applied [6] in predicting stock market prices in Nifty 50 in the Indian stock market. Out of many technical indicators, they optimized the parameters using the DPSO method, avoiding overfitting problems.

Cheng et al. [7] have proposed to estimate the possible outcome of a photovoltaic system which was influenced by several weather parameters. They proposed an ANFIS network whose parameters were tuned with the modified gray wolf optimization (MGWO) technique. They tried to combine many other techniques like ANFIS, MGWO, LUBE (lower upper bound estimation) and KDE (Kernel density estimation). Natalia et al. [8] applied the neuro-fuzzy inference system (NFIS) for rotor speed controller in wind turbines. They used a feed-forward neural network and used the gradient descent method for learning. Kabir and Kabir [9] compared the performances of different models that used different membership functions in the fuzzification layer of the ANFIS. Generally, we use fuzzy membership functions to map fuzzy values into crisp output using functions like triangular, trapezoidal, Gaussian, generalized bell, etc. Out of these, the trapezoidal function outperforms others when used with backpropagation and hybrid training methods. This is well known that RMSE is minimized with an increasing number of epochs.

Safiullah [10] demonstrated the application of ANFIS in the domain of car modeling, for example, quarter car model, where the vibration of the suspensor was calculated, and hence, its physical design parameters were set. Sasan et al. [11] introduced a hybrid mechanism that found the optimized parameters using emperors penguin colony (EPC) algorithms for ANFIS. They achieved a slightly better performance compared to classical evolutionary algorithms with standard classical datasets of body fat percentage, breast cancer Wisconsin, housing price prediction, etc. To optimize features or attributes for a dataset, many metaheuristic methods are used. A few of them are already mentioned in this section. Talpur et al. [12] reviewed deep-neuro-fuzzy-system (DNFS) from an exhaustive list of publications. They also focused on various applicability of DNFS in the field of image processing, distributed systems, health care operations, etc. Talpur et al. [13] used one simpler method, arithmetic optimization algorithm (AOA) that was applied to datasets with fewer attributes. They compared their results with DNFS and AOA. They found that for certain datasets, AOA performed better.

In [14], the authors used an ANFIS model which predicted the landslide patterns in a certain region of Iran. This geo-physical problem set used four classical optimization problems for feature selection, namely genetic algorithm (GA), PSO, differential evolutionary (DE) algorithm and ant colony optimization (ACO). For that spatial dataset, DE algorithm performed better. In [15], the authors chose bio-inspired algorithms, namely advanced binary ACO and CRPSO. Since support vector machine (SVM) is a feature selection technique from the high dimensional dataset and is supervised in nature, we can't apply it in unsupervised classification problems. They tested on the colon and prostate datasets to find an optimized feature set. Their finding showed that Adaptive binary particle swarm optimization (ABPSO)performed better than crazy PSO. But crazy PSO is a good algorithm for finding quality attributes. However, Shahbazi and Amiri [16] solved some frequency and voltage-level maintenance problems.

ADAPTIVE NEURO-FUZZY INFERENCE SYSTEM

Usually, FIS cannot learn anything from the previous dataset. It can only be estimated from an existing dataset. On the other hand, ANN can handle large datasets where lots of input-output are available for training and testing. However, ANN couldn't handle uncertainties and vagueness, which are often related to humanoid thoughts. There are a number of ways to build neuro-fuzzy systems, like FALCON, FUN, SONFIN, NFN, FINEST, GARIC and ANFIS. ANFIS is a hybridization of ANN and Takagi-Sugeno

FIS, that is, an FIS that uses ANN for data processing. It is a supervised learning method proposed by Jyh Shing Roger Jang and Chuen Tsai Sun in 1993. Its inference system uses "if-then-else" logic that learns how to approximate nonlinear functions. Even it is known as a universal estimator, an ANFIS can be used to optimize parameters that are obtained from a GA. ANFIS with GA can be best utilized in an intelligent energy management system in microgrids.

ANFIS architecture consists of several layers like an ANN. The first layer is an input layer, and the second layer is used for fuzzification purposes. In layer-2, we have to decide the membership function that is to be fed. During fuzzification, premise parameters are tuned in the training process. In layer-3, rule base is defined. Fuzzified values from layer-2 are taken as the input to layer-3 and used for making decisions. Essentially, layer-2 and layer-3 are fully connected networks of neurons. Number of neurons present in layer-3 depends on the number of fuzzified values in layer-2, and this is combinatorial in nature. Layer-4 and layer-5 of an ANFIS are designated as a de-fuzzification layer. Usually, in layer-4, the same output membership function is used to map real output from the previous layer. Consequent parameters are estimated during the training process to make it adaptive. Finally, in layer-5, we may get a human-understandable value by means of the weighted average method. The ANFIS system has the capability of self-learning. Its complexity increases when the number of inputs exceeds 15 which essentially increases the volume of its rule base. The following architecture shows the details of the workings of an ANFIS model (Figure 4.1).

ANFIS model uses a hybrid learning approach to tune premise parameters and consequent parameters. In the forward pass, premise parameters are fixed while consequent parameters are updated using the least square

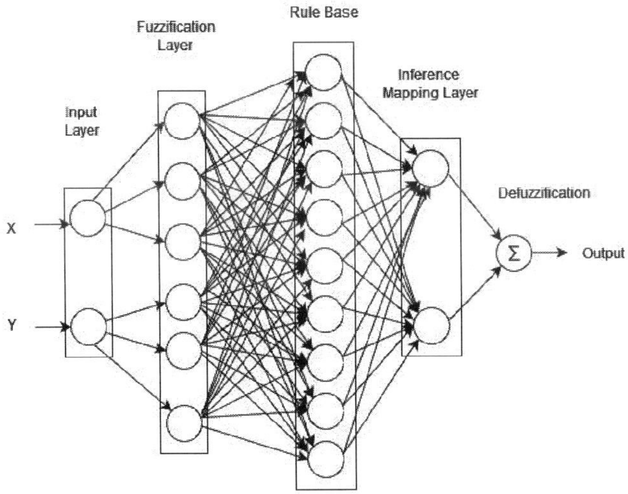

Figure 4.1 5-Layered ANFIS architecture.

estimate (LSE) method. In the backward pass, the consequent parameters remain unchanged, but premise parameters are updated using the gradient descent method. ANFIS is applied in time series prediction, decision-making, pattern recognition, data mining and controller design.

PARTICLE SWARM OPTIMIZATION (PSO)

Kennedy and Eberhart introduced this term in their publication "Particle Swarm Optimization" in 1995. PSO is a metaheuristic iterative process which chooses a candidate solution initially, and after several iterations, it converges to optimal solution(s). PSOs are attributed to the experiences of animal behavior like a school of fish or flock of birds while they are moving for food, habitat or reproduction. These algorithms have a drawback that we do not guarantee for global optimal solution. But, in reality, often it happens. It is convenient to use PSOs in multi-dimensional vector space to find global maxima or global minima.

Let in *i*th iteration, a particle P is located at a point $X_i\ [x_i(t),\ y_i(t)]$. This particle has a velocity $V_i(t)= [V_x^{\,i}(t),\ V_y^{\,i}(t)]$ for $i = 0,1,2,3,\ldots$, etc.

Now, in the next iteration, the particle gets a new position at $P(t + 1)$ and that position will be calculated as,

$$X^i(t+1) = X^i(t) + V^i(t) \tag{4.1}$$

or equivalently,

$$x^i(t+1) = x^i(t) + V_x^{\,i}(t+1) \tag{4.2}$$

$$y^i(t+1) = y^i(t) + V_y^{\,i}(t+1) \tag{4.3}$$

also, the velocity is updated by the following equation:

$$V^i(t+1) = w * v^i(t) + c_1 r_1\left(P_{\text{best}}{}^i - x^i(t)\right) + c_2 r_2\left(G_{\text{best}} - x^i(t)\right) \tag{4.4}$$

where r_1 and r_2 are random numbers between 0 and 1; constants w, c_1 and c_2 are parameters to the PSO algorithm; $P_{\text{best}}{}^i$ is the position that gives the best value ever explored by particle; and G_{best} is that explored by all the particles in the swarm. Using these equations iteratively, we can evaluate the best possible optimal (minima or maxima) for a function $f(x)$. PSO algorithm can be executed in parallel. However, they have not assured divergence; this algorithm can be applied to single-value optimum problems. In fact, real-life problems based on socioeconomic scenarios are successfully solved by this algorithm.

CRAZINESS-BASED PARTICLE SWARM OPTIMIZATION (CRPSO)

This approach was introduced by Roy and Ghoshal in 2008. CRPSO is a modified version of PSO where we search for an optimal solution close to the global optima. In the previous section, we have discussed the PSO where the velocity of a particle is described by equation (4.4). In CRPSO, craziness parameters are included in the velocity equation. Craziness is assumed to be a random behavior that may be followed by a swarm while moving with its group. It states that when a swarm moves toward a destination in search of food or habitat, some of the particles may show random movement. By keeping track of the random particles, new positions and velocity of the swarm are modified as

$$V^i(t) = r_2\text{sign}(r_3)v^i(t-1) + (1-r_2)\ c_1r_1\left(p_{\text{best}}{}^i(t-1) - s^i(t-1)\right)$$
$$+(1-r_2)c_2(1-r_1)\left(g_{\text{best}}(t-1)s^i(t-1)\right) \quad (4.5)$$

where $\text{sign}(r_3) = -1$ when $r_3 \leq 0.05$
and $\text{sign}(r_3) = 1$ when $r_3 > 0.05$.

The velocity of a particle is modified with craziness as follows,

$$V^i(t) = v^i(t) + P(r_4)\text{sign}(r_4)V^i{}_{\text{craziness}} \quad (4.6)$$

where $P(r_4) = 1$ when $r_4 \leq P_{\text{cr}}$
and $\text{sign}(r_4) = -1$ when $r_4 \leq 0.5$

The next position is updated based on equation (5.3).

$$S^i(t) = S^i(t-1) + v^i(t) \quad (4.7)$$

Equation (4.6) describes the velocity of a swarm, and equation (4.7) describes the position of a swarm with respect to its previous position. Iteratively, the positions and velocities are calculated which are known as epochs. In *i*th iteration, $p_{\text{best}}{}^i$ shows the local best solution, and at the end of the specified number of epochs, say n, we will get the global optimal value $g_{\text{best.}}$

In infinite impulse response (IIR) filter design, CRPSO produces better results in multi-dimensional search space compared to classical PSO and GA.

RECURSIVE LEAST SQUARE (RLS) ALGORITHM

German mathematician Gauss introduced this concept, which was used extensively in adaptive filter design for noisy signals. This method iteratively minimizes a weighted linear regression function to find optimal weights. This method has a faster convergence rate than its stochastic variants. RLS is an efficient method for learning the neuro-fuzzy system compared with other conventional methods. Here, we try to find the auto-correlation between two input parameters with their complex conjugate.

Let a signal $d(n)$ be transmitted over a noisy channel and the output is coming in the form of $d'(n)$with some additive noise $e(n)$, that is,

$$e(n) = d'(n) - d(n)$$

With a feedback mechanism, the error is recalculated as follows,

$$e(n) = e(n-1) + \Delta e(n-1).$$

Let there be a vector $Y[y_1,y_2,...,y_n]$ for a certain input to the system and we calculate its estimate $X'[x_1,x_2,...,x_n]$. There would be certain parameters $C[c_1,c_2,...,c_n]$ upon which the next value can be estimated. In this algorithm, the initial error is calculated, and this error value is repeatedly propagated to the next higher order iterations. The estimated update is given by,

$$X'_k = X'_{k-1} + K_k\left(y_k - c_k x'_{k-1}\right) \tag{4.8}$$

where K_k is the gain matrix which is calculated as,

$$K_k = P_{k-1}C_k{}^T\left(R_k + P_{k-1}C_kC_k{}^T\right) \tag{4.9}$$

and the estimation of the propagation error is given as,

$$P_k = (I - K_kC_k)P_{k-1}(I - K_kC_k)^T + K_kR_kK_k{}^T \tag{4.10}$$

or,

$$P_k = (I - K_kC_k)P_{k-1}. \tag{4.11}$$

NEURO-FUZZY CONVEX OPTIMIZATION

Convex programming constitutes a class of mathematical programming models where the objective function is convex and the constraints are formed by convex functions over a convex decision space. Convex optimization problems either reveal minimizing convex functions over a convex set or maximizing concave functions over a concave set.

The standard form is

$$\begin{aligned} \underset{x}{\text{minimize}} \quad & f(x) \\ \text{Subject to:} \quad & g_i(x) \le 0. \\ \text{and} \quad & h_i(x) = 0 \end{aligned} \tag{4.12}$$

Well-known mathematical optimization problems like linear programming, quadratic programming, second-order cone program, semi-definite programming and cone optimization are known as convex optimization.

There are many approaches to map a convex optimum problem into a fuzzy optimum system (Silva et al. [17]). These are Tang and Wang approach; Liu approach; Ammar and Khalifa approach; Canestrelli, Giove and Full´er approach; Tonon and Bernardini approach; extended Verdegay's linear approach, etc. Out of them, Tang and Wang approach is popular and is applied in a wide range of applications for transforming a convex optimization into fuzzy convex optimization. According to Tang and Wang's approach, the fuzzy transformation of objective function and constraints are transformed as follows:

$$\begin{array}{ll} \overline{\min} & \mathbf{c}^{t}\mathbf{x} + \frac{1}{2}\mathbf{x}^{t} \cdot \mathbf{Q}\mathbf{x} \\ \text{subject to:} & \mathbf{A}\mathbf{x} \leq \bar{b} \\ & \mathbf{x} \in \sigma \end{array} \quad . \tag{4.13}$$

A membership function for each fuzzy component of vector $\bar{b}$ is needed to solve this problem. The membership function is defined as follows:

$$\begin{aligned} \mu_i(y) &= 1; & y \leq b_i \\ &= 1 - \frac{y - bi}{di}; & b_i \leq y \leq b_i + d_i \\ &= 0; & y > b_i + d_i \end{aligned} \tag{4.14}$$

where d_i, $(i = 1,\ldots, m)$ is the allowed maximum tolerance in each restriction. Each fuzzy parameter in equation (13) can be defined by membership functions such that: $\mu_i : \mathrm{R} \to [0,1], \textit{for } i = 1,\ldots,m.$

With fuzzy logic and soft computing techniques have been used to simulate complicated and nonlinear decision problems. Partially defined or ill-defined objective functions as well as cost functions can be mapped to a fuzzy set. Thus, a nonlinear, complex and ill-defined convex problem is modeled into a fuzzy-based system. When a problem is mapped to fuzzy-based system, we may use a neural network (NN) model as described in the Adaptive neuro-fuzzy inference system section. In addition to this, if there are filled-in probabilistic data in these vectors and we have to map them to a fuzzy set, then Canestrelli, Giove and Full´er approach; Tonon and Bernardini approach; extended Verdegay's linear approach, etc. can be used.

RELATION BETWEEN DEEP NEURAL NETWORK (DNN) AND FUZZY-BASED SYSTEM

It is worth noting that both the deep neural network (DNN) and fuzzy-based systems are approaches in artificial intelligence and machine learning. Neural networks with multiple layers, known as DNN, mimic the structure and functions of the human brain. Each neuron calculates some output based on some input provided to it. The error generated from the output terminal is minimized by adjusting weights using gradient descent methods.

In contrast, fuzzy-based systems are the extension of Boolean systems where input values are imprecise. For example, if the temperature is represented as hot, cold, pleasant, comfortable, etc., then we couldn't just assign Boolean values 0 or 1, rather we may assign values between 0 and 1 for each condition. Essential steps involved in fuzzy-based systems are fuzzification, inference engine, rule base and defuzzification.

DNNs are capable of learning complex patterns from image data, voice/speech signals, live captured data from sensors, game playing and draw sketches, etc. DNN can extract important feature(s) from a large dataset without manual intervention. Whereas fuzzy-based systems are capable of handling uncertain and imprecise data. Human readable rules are mapped into if-then-else logic set from where inferences can be made. Fuzzy systems are useful in controller design, expert system (ES) design, designing of decision support system (DSS) and in robotics.

DNN and fuzzy-based systems are distinct in nature. However, we are trying to integrate them to implement autonomous systems (AS), which are useful in industrial automation. This integration is known as neuro-fuzzy-system (NFS) or fuzzy-neural networks (FNN). NFS uses the advantages of learning complex patterns from a large dataset as well as the strong logical reasoning capability of fuzzy systems. Obviously, FNNs can tune the rule set from a large dataset and improve the performance of industrial problems with greater adaptability.

PERFORMANCE COMPARISON AMONG PSO, CRPSO AND RLS ALGORITHMS

RLS algorithm is an adaptive algorithm which works with complex datasets with greater reliability. It is very useful in Kalman filter design in signal processing tasks and in control system design. One of its variant, recursive mean square (RMS) is the simplest form of adaptive algorithm, but it offers a better convergence rate. RLS minimizes the cost function to find the best parameter settings but is not directly comparable with PSO or CRPSO in the context of rate of convergence. PSO is sometimes stuck into local optima, whereas CRPSO is more robust than PSO since it considers all the randomness of the participating parameters. CRPSO is computationally more complex than PSO and RLS but can produce better results.

In electrical systems, PSO performs optimization tasks like minimization of power losses, optimized placement of power generating stations, economic dispatch of load in real time, load forecasting, operational costs, etc. Its convergence rate is good as mentioned earlier. The performance of CRPSO is quite complex in nature but more accurate than PSO since its learning mechanism is based on the randomness of moving swarms. CRPSO can also be used in multi-objective optimization. Whereas RLS is not directly associated with optimization, rather it performs parameter estimation and tuning of effective parameters.

CONCLUSION

In ANFIS, it was observed that DPSO performs better in reducing dimensions. Its performance was compared with well-known error measurements like RMSE, MAAPE, Theil's U, etc. In the fuzzification layer, the trapezoidal membership function performs well with a minimum of 50 epochs. Fuzzy membership functions are also tested to get optimized power usage in a vehicle-to-home solar power system. To get better performance, optimization of fuzzy rules is also necessary. Self-tuned or adaptive neurons can be used in PID controller design, but it can deal with a limited number of controlling parameters. In the case of high dimensional data, we may adopt many algorithms like clustering-based population creation with binary gravitational search [17], PSO, etc.

The applicability of neuro-fuzzy systems in electrical systems is very popular: from power generation forecasting to usage forecasting, optimizing photovoltaic energy systems, making PID controllers of different turbines and determining load factors in hybrid grids where conventional energy is amalgamated with renewable energy sources and many more. Hence, we may conclude that in predicting power generation and usage, the NFIS plays a key role.

REFERENCES

1. Mahdi Shafaati Shemami, Mohammad Saad Alam, M. S. Jamil Asghar, Samir M. Shariff, 'Adaptive Neuro-Fuzzy Inference System (ANFIS) for Optimization of Solar Based Electric Vehicle-to-Home (V2H) Fuzzy Inference System (FIS) Controller', ISSN: 978-1-5386-9310-0, IEEE, 2019.
2. Appolo Tankeh, E. II. Mamdaiii, 'Convex Fuzzy Controller: Neuro-Fuzzy and Convex Optimisation', ISSN: 0-71103-3796-4, IEEE, 1997.

3. Ola Abdelnasser Mosaad Ali, Helmy Mohamed El-Zoghby, Abdel Ghany Mohamed Abdel Ghany, 'Maximizing the Generated Power from Hybrid Wind-Solar System Based on Fuzzy Self Tuning Single Neuron PID Controller', in *2018 Twentieth International Middle East Power Systems Conference (MEPCON)*, ISSN:978-1-5386-6654-8, IEEE, 2018.
4. Ritam Guha, Manosij Ghosh, Akash Chakrabarti, Ram Sarkar, Seyedali Mirjalili, 'Introducing clustering based population in Binary Gravitational Search Algorithm for Feature Selection', *Applied Soft Computing Journal*, 2020. https://doi.org/10.1016/j.asoc.2020.106341.
5. Jeffrey Andrew-Cotter, M. Nasir Uddin, Ifte Khairul Amin, 'Particle Swarm Optimization based Adaptive Neuro-Fuzzy Inference System for MPPT Control of a Three-Phase Grid-Connected Photovoltaic System', ISSN:978-1-5386-9350-6, IEEE, 2019.
6. Gourav Kumar, Sanjeev Jain, Uday Pratap Singh, 'Neuro-Fuzzy and Particle Swarm Optimization based Hybrid Approach for Stock Price Forecasting', in *International Conference on Emerging Smart Computing and Informatics (ESCI)*, ISSN: 978-1-7281-8519-4, IEEE, 2021.
7. Lilin Cheng, Haixiang Zang, Tao Ding, Miaomiao Wang, Zhinong Wei, Guoqiang Sun, 'A Combined Optimization Structure of Adaptive Neuro-fuzzy Inference System for Probabilistic Photovoltaic Power Forecasting', ISSN: 978-1-7281-1981-6, IEEE, 2019.
8. Natalia V. Zubova, Andrey A. Achitaev, 'Application of Neuro-Fuzzy Control Systems for Increasing the Energy Efficiency of Wind Turbines', in *14th International Scientific-Technical Conference APEIE*, ISSN: 978-1-5386-7054-5, 2018, IEEE.
9. Monika Kabir, Mir Md. Jahangir Kabir, 'Fuzzy Membership Function Design: An Adaptive Neuro-Fuzzy Inference System (ANFIS) Based Approach', in *International Conference on Computer Communication and Informatics (ICCCI-2021)*, ISSN: 978-1-7281-5875-4, IEEE, 2021.
10. Mohammad Adom Safiullah, 'Modeling of Optimized Neuro-Fuzzy Logic Based Active Vibration Control Method for Automotive Suspension', Masters Theses, Grand Valley State University, https://scholarworks.gvsu.edu/theses/921, December, 2018.
11. Sasan Harifi, Madjid Khalilian, Javad Mohammadzadeh, and Sadoullah Ebrahimnejad, 'Optimizing a Neuro-Fuzzy System Based on Nature-Inspired Emperor Penguins Colony Optimization Algorithm', *IEEE Transactions on Fuzzy Systems*, vol. 28, no. 6, 2020. doi:10.1109/TFUZZ.2020.2984201,
12. Noureen Talpur, Said Jadid Abdulkadir, Hitham Alhussian, Mohd Hilmi Hasan, Norshakirah Aziz, Alwi Bamhdi, 'Deep Neuro-Fuzzy System Application Trends, Challenges, and Future Perspectives: A Systematic Survey', *Artificial Intelligence Review*, 2023. https://doi.org/10.1007/s10462-022-10188-3.
13. Noureen Talpur, Said Jadid Abdulkadir, Hitham Alhussian, Mohd Hilmi Hasan, Mohd Hafizul Afifi Abdullah, 'Optimizing Deep Neuro-Fuzzy Classifier with a Novel Evolutionary Arithmetic Optimization Algorithm', *Journal of Computational Science*, 2022. https://doi.org/10.1016/j.jocs.2022.101867.

14. Mohammad Mehrabi, Biswajeet Pradhan, Hossein Moayedi, Abdullah Alamri, 'Optimizing an Adaptive Neuro-Fuzzy Inference System for Spatial Prediction of Landslide Susceptibility Using Four State-of-the-art Metaheuristic Techniques', 2020. https://doi.org/10.3390/s20061723.
15. Neni Alya Firdausanti, Irhamah, 'Advanced Binary Ant Colony Optimization for Feature Selection on High-Dimensional Data', *Procedia Computer Science* 161, 2019, pp. 638–646.
16. Mohamadmahdi Shahbazi, Farhad Amiri, 'Designing a Neuro-Fuzzy controller with CRPSO and RLSE algorithms to control voltage and frequency in an isolated microgrid', in *34th International Power System Conference (PSC2019)*, ISSN: 978-1-7281-5273-8, IEEE, 2019.
17. Ricardo C. Silva, Carlos Cruz, Jos´e L. Verdegay, Akebo Yamakami, '*A Survey of Fuzzy Convex Programming Models*', pp. 127–143, Springer, Berlin, 2010.

Chapter 5

OPF incorporating FACTS tools with renewable energy by GWA

Tushnik Sarkar, Susanta Dutta, Provas Kumar Roy, Sunanda Hazra, Chandan Paul, Sourav Paul, and Sneha Sultana

INTRODUCTION

The power system is one of the most complex networks that provides the demanding load to the different types of consumers according to their various types of needs as per their time. They must deal with a severe condition where they are overloaded because we have limited transmission and generation capacity, and the condition gets worse under different types of contingency conditions. The OPF acts as a tool, and its key objective is to diminish total production costs, load demand accomplishment, and operation of devices within their lowest and highest limits.

Both conventional and evolutionary optimization techniques that have been discussed in the literature conventional techniques have a problem of identification of global minimum, necessitating the use of evolutionary optimization techniques like the solution of OPF using oppositional Krill Herd algorithm [1]. In Ref. [2], the hybrid strategy of backtraping-assisted elephant herding optimization (BAEHO) is used to address the optimization issue. Grey wolf approximation (GWA) with populace-based update evaluation (GW-PUE) algorithm is found in Ref. [3] to attain OPF solution. The problem of sub-optimality was resolved by the algorithm in Ref. [4]. In Ref. [5], fitness-distance balance-based adaptive guided differential evolution algorithm was presented to solve the OPF problem involving renewable sources [6]. The idea is to view the Joint Chance Constrained (JCC)-OPF problem as a distribution matching problem. The Maximum Mean Discrepancy (MMD) quantifies the distance between two probability distributions embedded into reproducing kernel Hilbert space (RKHS) and thus provides an efficient way to minimize the distance between distributions [7]. Major contributions with respect to previous works include the consideration of all types of contingencies and control actions, the leveraging of the methodology to the Security Constrained Optimal Power Flow (SCOPF) setting, including multiple suppression mentalities for identifying control actions, and [8] the further showcasing of the academia-industry gap in a topic that is relatively obscure in academia but crucial in industry. Sheikhan et al. [9]

 DOI: 10.1201/9781003628279-5

proposed a new alternative optimization objective that may be more suitable for Intentional Controlled Islanding (ICI) than the minimization of load shedding, a new set of island connectivity constraints, a new set of constraints for direct current (DC) OPF with switching, and a new mixed-integer linear programming (MILP) heuristic to find initial feasible solutions for ICI [10]. A novel algebra is developed to represent the problem of OPF as an optimization problem for the proposed coreset definition. A novel coreset construction algorithm that can approximate the OPF solutions is subsequently proposed in order to improve the OPF construction speed [11]. The present approach relies on solving an OPF locally in each alternating current (AC) agent and AC-DC agent that only requires exchanging the voltages corresponding to adjacent nodes. In this chapter, the Grey wolf optimization algorithm is employed for OPF problems, including renewable energy, along with the inclusion of TCSC and TCPS, and the result obtained is compared with Krill Herd Algorithm (KHA) [1].

MODELING OF FACTS DEVICES

Modeling of TCSC

A series-connected FACTS controller termed as TCSC may rapidly change the transmission network's impedance by altering the firing angle under different operating situations. The static form of the network with TCSC attached between *m*th and*n*thbus is shown in Figure 5.1. The active power flow in the compensated line can be controlled and maintained over a wide range of operating conditions.

The equation of power flow for the branch with TCSC can expressed as follows [13]:

$$Q_{mn} = -V_m^2 B_{mn} - V_m V_n G_{mn} \sin(\partial_m - \partial_n) + V_m V_n B_{mn} \cos(\partial_m - \partial_n) \tag{5.1}$$

$$P_{mn} = V_m^2 G_{mn} - V_m V_n G_{mn} \cos(\partial_m - \partial_n) - V_m V_n B_{mn} \sin(\partial_m - \partial_n) \tag{5.2}$$

$$Q_{mn} = -V_m^2 B_{mn} + V_m V_n G_{mn} \sin(\partial_m - \partial_n) + V_m\ V_n B_{mn} \cos(\partial_m - \partial_n) \tag{5.3}$$

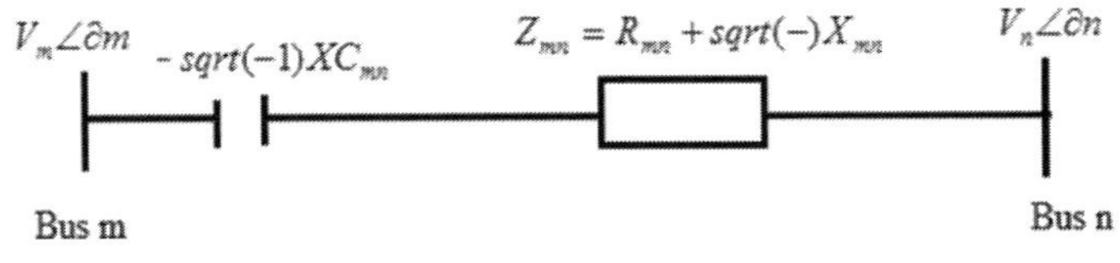

Figure 5.1 Model of transmission line with TCSC.

$$(G_{mn}) = R_{mn} / \{R_{mn}^2 + (X_{mn} - X_{C_{mn}})^2\} \tag{5.4}$$

$$B_{mn} = (X_{mn} - X_{C_{mn}}) / \{R_{mn}^2 + (X_{mn} - X_{C_{mn}})^2\} \tag{5.5}$$

where the transmission line's resistance and reactance are R_{mn} and X_{mn}, respectively, which is connected between mth and nth bus and the TCSC placed there having reactance XC_{mn}.

TCPS model

A phase-shifting transformer with a tapping ratio of $1:1\angle\phi$ and a series admittance of $Y_{mn} = (G_{mn} - jB_{mn})$ may be utilized to model TCPS as shown in Figure 5.2.

The power flow equation of the line having TCPS connected between mth and$n$th buses can be derived as follows [4,9]:

$$P_{mn} = \frac{V_m^2 G_{mn}}{\cos^2\varphi} - \frac{V_m V_n}{\cos\varphi}[G_{mn}\cos(\partial_m - \partial_n + \varphi) + B_{mn}\sin(\partial_m - \partial_n + \varphi)] \tag{5.6}$$

$$Q_{mn} = -\frac{V_m^2 B_{mn}}{\cos^2\varphi} - \frac{V_m V_n}{\cos\varphi}[G_{mn}\sin(\partial_m - \partial_n + \varphi) - B_{mn}\cos(\partial_m - \partial_n + \varphi)] \tag{5.7}$$

$$P_{nm} = V_n^2 G_{mn} - \frac{V_m V_n}{\cos\varphi}[G_{mn}\cos(\partial_m - \partial_n + \varphi) - B_{mn}\sin(\partial_m - \partial_n + \varphi)] \tag{5.8}$$

$$Q_{nm} = -V_n^2 B_{mn} + \frac{V_m V_n}{\cos\varphi}[G_{mn}\sin(\partial_m - \partial_n + \varphi) + B_{mn}\cos(\partial_m - \partial_n + \varphi)] \tag{5.9}$$

The injected TCPS reactive and active powers at mth and nth bus can be derived as follows [13]:

$$P_{ms} = -G_{mn}V_m^2\tan^2\phi - V_m V_n\tan\phi[G_{mn}\sin(\partial_m - \partial_n) - B_{mn}\cos(\partial_m - \partial_n)] \tag{5.10}$$

$$Q_{ms} = B_{mn}V_m^2\tan^2\phi + V_m V_n\tan\phi[G_{mn}\cos(\partial_m - \partial_n) + B_{mn}\sin(\partial_m - \partial_n)] \tag{5.11}$$

where Q_{ms}, P_{ms} are the injected reactive and active power, respectively, of TCPS at the mth bus.

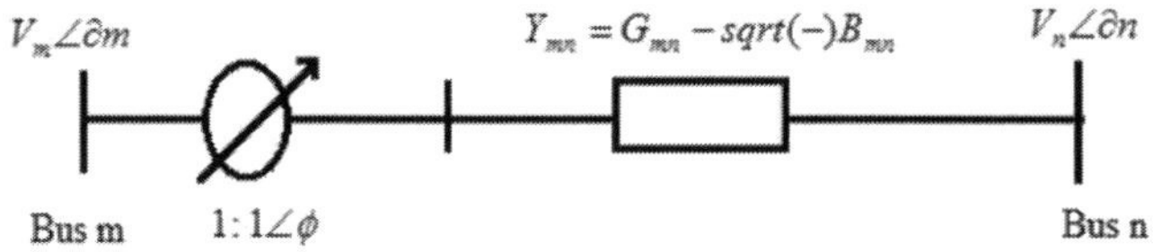

Figure 5.2 Model of transmission line with TCPS.

$$P_{ns} = -V_m V_n \tan\phi[G_{mn}\sin(\partial_m - \partial_n) + B_{mn}\cos(\partial_m - \partial_n)] \tag{5.12}$$

$$Q_{ns} = -V_m V_n \tan\phi[G_{mn}\cos(\partial_m - \partial_n) - B_{mn}\sin(\partial_m - \partial_n)] \tag{5.13}$$

MODELING OF RENEWABLE ENERGY

Modeling of wind power generation

The uncertainty present in wind speed creates problems for incorporating the wind power with the existing network. The Weibull [14] PDF is commonly preferred to express wind pace variability, which is provided in (5.14):

$$f_m(v_w) = \frac{l}{c}\left(\frac{v_w}{c}\right)^{l-1} \times e^{-\left(\frac{v_w}{c}\right)^l} \tag{5.14}$$

where v_w indicates wind present velocity; m stands for random wind number; and $c>0$ and $l>0$ depict scale factor and the subsequent shape factor, respectively. The cumulative density function (CDF) is given by (5.15):

$$F_m(v_w) = 1 - e^{-\left(\frac{v_w}{c}\right)^l} \tag{5.15}$$

The wind power is evaluated as follows:

$$P_w = \begin{cases} 0 & v_w < v_{\mathrm{IN}} \text{ or } v_w > v_{\mathrm{OUT}} \\ \dfrac{P_{wr}(v_w - v_{\mathrm{IN}})}{v_R - v_{\mathrm{IN}}} & v_{\mathrm{IN}} \le v_w < v_R \\ P_{wr} & v_R \le v_w < v_{\mathrm{OUT}} \end{cases} \tag{5.16}$$

where v_R represents the rated wind velocity, v_{IN} is cut-in wind velocity, v_{OUT} indicates cut-out velocity, P_{wr} is rated wind power, and P_w is output wind power.

Now, differing wind speed zones' possibility of wind energy can be explained by:

$$f(P_w)_{|P_w=0} = 1 - e^{-\left(\frac{v_{\mathrm{IN}}}{c}\right)^t} + e^{-\left(\frac{v_{\mathrm{OUT}}}{c}\right)^t} \tag{5.17}$$

$$f(P_w)_{|P_w=P_{wr}} = e^{-\left(\frac{V_R}{c}\right)^t} - e^{-\left(\frac{v_{\mathrm{OUT}}}{c}\right)^l} \tag{5.18}$$

$$f_{P_w}(P_w) = \begin{cases} 0 & P_w < 0 \\ \dfrac{lu}{cP_{wr}} \left(\dfrac{v_{IN} + u\dfrac{P_w}{P_{wr}}}{c} \right)^{l-1} \times e^{-\left(\frac{v_{IN} + u\frac{P_w}{P_{wr}}}{c} \right)^{l}} & 0 \le P_w < P_{wr} \\ 1 & P_w \ge P_{wr} \end{cases} \tag{5.19}$$

Wind power cost calculation

The placement of wind power generating units into the grid at times when electricity is needed will be impacted by wind uncertainty. Power generation is unreliable due to wind speed fluctuation. The Weibull-probability density function will be used to analyze the anticipated uncertainty costs of wind power. Overestimation and underestimation are used to define this function (5.8):

$$\mathrm{TC}_w \sum_{l=1}^{N_w} C_{wl}(P_{wl}) = \sum_{l=1}^{N_w} \left(C_{wl}^{O} + C_{wl}^{U} \right) \tag{5.20}$$

where TC_w stands for the total cost of wind and N_w for the total number of wind generators.

Modeling of solar power generation

The probability distribution of solar irradiance i_R is expressed by [14]:

$$f_{pv}(i_R) = \frac{1}{i_g d\sqrt{2\pi}} e^{\frac{-(ini_g - M)^2}{(2d)^2}} \quad \text{for } i_R > 0 \tag{5.21}$$

The solar unit's power output is expressed as follows:

$$\begin{cases} P_s = P_{sr}\left(\dfrac{i_R{}^2}{i_{\mathrm{Rsd}} R_C} \right) & \text{for } 0 < i_R < R_C \\ \quad\; = P_{sr}\left(\dfrac{i_R}{i_{\mathrm{Rsd}}} \right) & \text{for } i_R > R_C \end{cases} \tag{5.22}$$

where sr is the rated power of solar unit, S is the output power of the solar unit, i_{Rsd} (= 1000 W/m^2) implies standard irradiance of solar cell, and R_C (= 150 W/m^2) is a specific irradiance point.

Cost function: solar energy

The cost of producing electricity from a solar panel is calculated by adding together three separate cost functions.

$$C_{sl}(P_{sl}) = C_{sl}^{d} + C_{sl}^{O} + C_{sl}^{U} \tag{5.23}$$

where C_{sl}^{d}, C_{sl}^{O} and C_{sl}^{U} are the direct cost and underestimation cost of the lth solar unit, respectively.

Determination of direct cost of solar

The cost of producing solar power is referred to as direct costs. This word is not applicable if the system operator owns the solar farm. The following equation determines the direct cost of solar energy:

$$C_{sl}^{d} = d_{l}^{s} P_{sshl}, \quad \text{where } l = 1,2,3,\ldots,ns \tag{5.24}$$

Here, d_{l}^{s} represents the direct cost coefficients for the lth solar unit and P_{sshl} represents the schedule power of the lth unit.

Determination of overestimation cost of solar

The overestimation cost is determined using the formula below if the amount of solar power available is less than the amount of electricity scheduled.

$$\begin{cases} C_{sl}^{O} = \mathrm{PF}_{sl}^{O}(P_{sshl} - P_{savl}) \\ \quad = \mathrm{PF}_{sl}^{O} \displaystyle\int_{0}^{P_{ssl}} (P_{sshl} - P_{s}) f_{P_s}(P_s)\, dP_s' \end{cases}$$

MATHEMATICAL PROBLEM FORMULATION

Objective function

Minimization of fuel cost

The following table [10,11] provides the total fuel cost function for generating units with quadratic cost functions and without taking the valve point discontinuity into account.

$$\mathrm{FC} = \sum_{m=1}^{\mathrm{NG}} \left(a_m P_{Gm}^{\ 2} + b_m P_{Gm} + c_m \right) \tag{5.25}$$

where FC is the total fuel cost of generating units, a_m, b_m, c_m are fuel cost coefficients of the mth generator, and NG is the number of generator buses. The following equation [10] provides the generation cost for producing units with valve point discontinuities.

$$FC = \sum_{m=1}^{NG} a_m P_{Gm}{}^2 + b_m P_{Gm} + c_m + \left| d_m \times \sin\left(e_m \times \left(P_{Gm}^{\min} - P_{Gm}\right)\right) \right| d_m, e_m \quad (5.26)$$

Fuel cost coefficient with valve point loading effect of mth generator is represented by a_m, b_m, c_m, d_m, e_m and NG is the number of generator buses.

Minimization of transmission losses

The mathematical expression for describing the total power loss considering G_x, that is, the conductance of the line connected between the mth and nth buses is as below [5,6]:

$$P_{\text{LOSS}} = \sum_{X=1}^{NL} G_x [V_m^2 + V_n^2 - 2|V_m||V_n| \cos(\partial_m - \partial_n)] \quad (5.27)$$

where NL is the number of transmission lines and P_{LOSS} is the total power loss.

Constraints

Equality constraints

The sets of equality constraints that reflect the load flow equations are listed below [4,9]:

$$\sum_{m=1}^{NB} (PG_m - PL_m) + \sum_{m=1}^{NTCPS} P_{ms} = \sum_{m=1}^{NB} \sum_{n=1}^{NB} |V_m||V_n||Y_{mn}| \cos(\theta_{mn} - \partial_{mn})| \quad (5.28)$$

$$\sum_{m=1}^{NB} (QG_m - QL_m) + \sum_{m=1}^{NTCPS} Q_{ms} = -\sum_{m=1}^{NB} \sum_{n=1}^{NB} |V_m||V_n||Y_{mn}| \sin(\theta_{mn} - \partial_{mn})| \quad (5.29)$$

where NTCSC is the number of (TCSC) and NTCPS is the number of (TCPS) used in the system.

PL_m and QL_m are the active and reactive power demand of the mth bus, respectively. Y_{mn} is the total admittance connected between the mth and nth bus, respectively, and θ_{mn} is the admittance angle of the transmission line connected between the mth and nth bus.

Inequality constraints

Generator Constraints Using [5,9]

$$\begin{cases} V_{Gm}^{\min} \le V_{Gm} \le V_{Gm}^{\max} \\ P_{Gm}^{\min} \le P_{Gm} \le P_{Gm}^{\max} & m \in \mathrm{NG} \\ Q_{Gm}^{\min} \le Q_{Gm} \le Q_{Gm}^{\max} \end{cases} \tag{5.30}$$

where $Q_{Gm}^{\min}$ and $Q_{Gm}^{\max}$ are minimum and maximum reactive power limits of the mth bus, respectively. $V_{Gm}^{\min}$ and $V_{Gm}^{\max}$ are minimum and maximum voltage limits of the mth generator, respectively. $P_{Gm}^{\min}$ and $P_{Gm}^{\max}$ are minimum and maximum active power limits of the mth bus, respectively.

Load Bus Constraints Using [5,9]

$$V_{Lm}^{\min} \le V_{Lm} \le V_{Lm}^{\max} \qquad m \in NL \tag{5.31}$$

where $V_{Lm}^{\min}$ and $V_{Lm}^{\max}$ are minimum and maximum voltage limits of the mth load bus, respectively.

Transmission Line Constraints Using [5,9]

$$S_{Lm} \le S_{Lm}^{\max} \qquad m \in NTL \tag{5.32}$$

where S_{Lm} is the apparent power flow limit of the mth branch and $S_{Lm}^{\max}$ is the maximum apparent power flow limit of the mth branch.

Transformer Tap Constraints Using [5,9]

$$T_{m}^{\min} \le T_{m} \le T_{m}^{\max} \qquad m \in NT \tag{5.33}$$

where $T_{m}^{\min}$ and $T_{m}^{\max}$ are minimum and maximum tap setting limits of the mth regulating transformer, respectively.

Shunt Compensator Constraints Using [5,9]

$$Q_{Cm}^{\min} \le Q_{Cm} \le Q_{Cm}^{\max} \qquad m \in NC \tag{5.34}$$

$Q_{Cm}^{\min}$ and $Q_{Cm}^{\max}$ are minimum and maximum reactive power injection limits of the mth compensator, respectively.

TCSC Reactance Constraints Using [5,9]

$$X_{tm}^{\min} \leq X_{tm} \leq X_{tm}^{\max} \qquad m \in NTCSC \tag{5.35}$$

where $X_{tm}^{\min}$ and $X_{tm}^{\max}$ are minimum and maximum reactance limits of the mth TCSC, respectively.

TCPS Phase Shift Constraints Using [5,9]

$$\alpha_{m}^{\min} \leq \alpha_{m} \leq \alpha_{m}^{\max} \qquad m \in NTCPS \tag{5.36}$$

$\alpha_{m}^{\min}$ and $\alpha_{m}^{\max}$ are minimum and maximum phase shift angles of the mth TCPS, respectively.

ALGORITHMS

Grey Wolf optimization

Mirjalili et al. have developed the innovative meta-heuristic optimization method known as "Grey Wolf optimization." Grey wolves are typically found in groups of 5–12 individuals and are regarded as the top predators in 2014 [12]. In the pack, the wolves are classified as alpha, beta, delta, and omega. This optimization approach largely focuses on the natural behavior of the grey wolves, their hunting mechanism, and their social leadership hierarchy. The fittest solution in the social hierarchy of grey wolves is alpha, which can be either male or female. Alpha is followed by beta, delta, and the remainder that are classified under omega, which always comes after alpha, beta, and delta. The alphas are responsible for leading the entire pack. In order for GWO to run, it is only necessary to specify the population size and the maximum number of iterations. The population size in GWO is determined by the number of grey wolves, and the fitness value is determined by the wolf's distance from its prey. The biggest obstacle is achieving a target population size. When identifying the global optimum solution, the GWO algorithm does not depend on the gradient of the objective function.

In a multi-dimensional search space, the wolves—alpha, beta, and delta—move in packs. The main variable in the GWO algorithm is each wolf's position, which it continuously updates to a better one. Thus, the wolf estimates the victim's location in the search space and updates their position at random intervals. Because of where the alpha, beta, and delta are in the search space, the final position is typically in a circle around the prey. The best outcome is kept in the GWO algorithm: the basic functions that the wolves carry out.

Searching for prey

The entire group splits off and explores the search area as widely as possible during the search or exploration phase. If the random result is larger than 1 or less than –1, the wolves diverge to the prey, and when the prey is located, the entire pack moves there.

The most crucial goal of the GWO algorithm is finding the prey. As can be seen from (5.22), a vector that provides a random weight for prey searching also impacts how far away the prey is from the hunter; therefore, using both of these vectors together makes it easier to explore a large area.

Encircling the prey

The following equations are taken into consideration when utilizing [12] the GWO algorithm for wolves to surround their prey:

$$\vec{D}N = \left| \vec{C}N \, \vec{X}_P(t) - \vec{X}(t) \right| \tag{5.37}$$

$$\vec{X}(t+1) = \vec{X}_P(t) - \vec{A}N \, \vec{D}N \tag{5.38}$$

Using [12,15], where t is the current iteration and $\vec{X}_P(t)$ represents the target's current position, A and C are constants.

$$\begin{aligned} \vec{A}N &= 2\vec{a} \times \vec{r}_1 - \vec{a} \\ \vec{C}N &= 2\vec{r}_2 \end{aligned} \tag{5.39}$$

where r_1 and r_2 are two random vectors between [0,1], and over each course of iteration, $\vec{a}$ decreases linearly from 2 to 0.

Hunting

Although beta and deltas also take part in the hunting process, the alphas direct the entire group in its pursuit of the prey. Since the precise location of the prey is still unknown, the wolves constantly update their positions to determine the best potential site for the prey. Three best options in terms of alpha, beta, and delta are stored. The hunting process often begins when the prey stops traveling in the circle defined by alpha, beta, and delta, even though the end position is likewise random. The mathematical formulas that follow describe the hunting phase and update where the wolves are located [12,15].

$$\vec{DN}_\alpha = \left|\vec{CN}_\alpha \times \vec{XP}_\alpha(t) + \vec{X}\right| \tag{5.40}$$

$$\vec{DN}_\beta = \left|\vec{CN}_\beta \times X\vec{P}_\beta(t) + \vec{X}\right|$$

$$\vec{DN}_\partial = \left|\vec{CN}_\partial \times X\vec{P}_\partial(t) + \vec{X}\right| \tag{5.41}$$

Finally, position was modified as using [12,15]

$$\vec{X}_{\alpha 1} = \vec{X}_\alpha - AN1\,\vec{DN}_\alpha \tag{5.42}$$

$$\vec{X}_{\beta 1} = \vec{X}_\beta - AN2\,\vec{DN}_\beta \tag{5.43}$$

$$\vec{X}_{\partial 1} = \vec{X}_\partial - \vec{AN}\,3\,\vec{DN}_\partial \tag{5.44}$$

$$\vec{X}(t+1) = \frac{\vec{X}_{\alpha 1} + \vec{X}_{\beta 1} + \vec{X}_{\partial 1}}{3} \tag{5.45}$$

Attacking prey (exploitation)

Based on the equations provided in the hunting phase (5.31), the alpha, beta, and delta wolves' locations in the GWO algorithm aid in improving the search agents' positions. Wolves cease hunting when their prey stops moving, which is primarily determined by their random values.

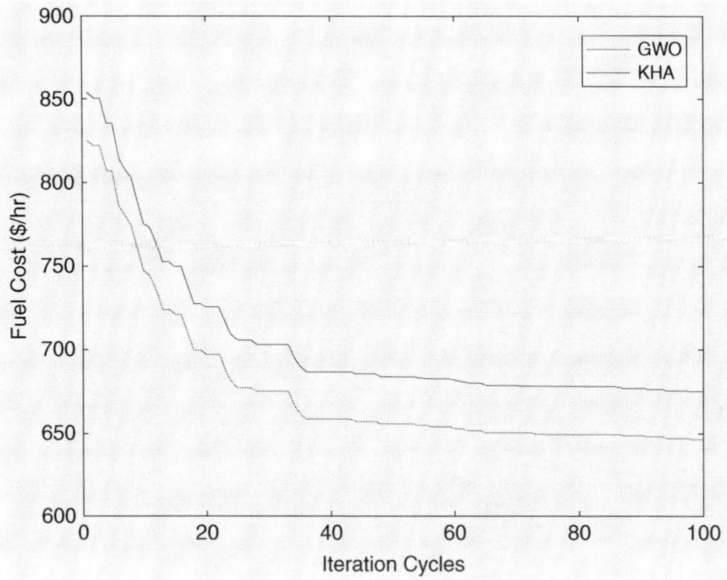

Figure 5.3 Cost convergence graph using GWO algorithm.

SIMULATION RESULTS AND DISCUSSIONS

Case study I: cost minimization

i. Without considering the valve point effect: the results for the proposed GWO algorithm and KHA [1] are presented in Table 5.1. The minimum fuel cost obtained from the proposed GWO algorithm is 674.0612 $/h, which is 1.4% less than the KHA [1] result. Figure 5.4 displays the fuel cost ($/h) convergence characteristics, demonstrating more uniform convergence of GWO.

ii. The results for OPF for the proposed algorithms, i.e., GWO and KHA [1], are shown in Table 5.2. Taking the valve point effect into account, the results show that the minimum fuel cost obtained from the proposed GWO algorithm is 648.0618 $/h, which is 4.24% less than the best-proposed result KHA [1]. The convergence profile of fuel cost ($/h) also demonstrates smoother convergence of GWO.

Table 5.1 Comparison of control variable settings and total fuel cost obtained by KHA and GWO (without valve point effect)

Algorithms	*KHA [1] with wind farm*	*GWO (with wind and solar farm) (proposed)*
PG_1 (MW)	137.526	138.5470
PG_2 (MW)	41.9122	21.5900
PG_5 (MW)	19.3311	29.7512
PG_8 (MW)	15.5927	15.7560
PG_{11} (MW)	20.9936	19.0312
PG_{13} (MW)	17.7110	21.1289
Total P_G (MW)	253.067	245.8043
P_{wind} (MW)	34.4126	20.0310
P_{solar} (MW)	0.0000	21.6820
$X_{C\,3\text{–}4}$ (PU)	–	0.0157
$X_{C\,19\text{–}20}$ (PU)	–	0.0214
φ_{5-7}(°)	–	–0.5504
φ_{10-22}(°)	–	–0.0251
Fuel cost ($/h)	683.646	674.0612
P_{LOSS} (MW)	4.08011	4.1173
CPU time (s)	188.100	187.154

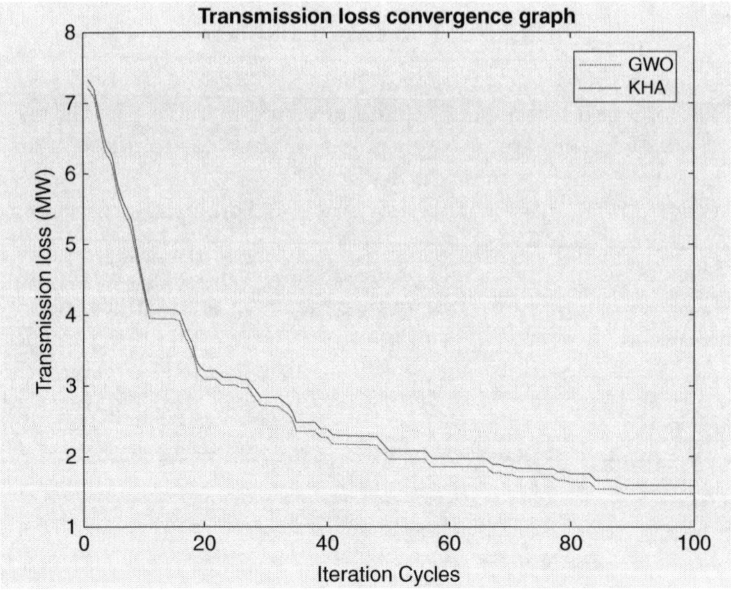

Figure 5.4 Transmission loss convergence graph using GWO algorithm.

Table 5.2 Best control variable settings for active power transmission loss minimization objective of modified IEEE-30 bus test power system offered by different algorithms

Algorithms	*KHA [1] with wind farm*	*GWO (with wind and solar farm) (proposed)*
PG_1 (MW)	15.7459	55.9800
PG_2 (MW)	80.0000	42.5500
PG_5 (MW)	50.0000	45.7800
PG_8 (MW)	35.0000	30.6700
PG_{11} (MW)	30.0000	25.8700
PG_{13} (MW)	40.0000	20.8700
Total P_G (MW)	250.745	221.72
P_{wind} (MW)	34.4212	33.2510
P_{solar} (MW)	0.0000	30.0000
$X_{C\,3-4}$ (PU)	–	0.0142
$X_{C\,19-20}$ (PU)	–	0.0029
φ_{5-7} (°)	–	–0.4226
φ_{10-22} (°)	–	–0.5434
Fuel cost ($/h)	918.639	733.692
P_{LOSS} (MW)	1.76710	1.5710
CPU time (seconds)	174.300	174.110

Case study 2: transmission loss minimization

Algorithms	*KHA [1] with wind farm*	*GWO (with wind and solar farm) (proposed)*
PG_1 (MW)	130.2460	146.3010
PG_2 (MW)	39.4530	22.0000
PG_5 (MW)	32.9689	16.0890
PG_8 (MW)	29.5335	30.0342
PG_{11} (MW)	11.8032	10.0230
PG_{13} (MW)	12.0000	12.0070
Total P_G (MW)	256.005	236.4542
P_{wind} (MW)	32.6020	30.2000
P_{solar} (MW)	0.0000	21.0000
$X_{C\,3-4}$ (PU)	–	0.0119
$X_{C\,19-20}$ (PU)	–	0.0248
φ_{5-7} (°)	–	–0.1801
φ_{10-22} (°)	–	0.2596
Fuel cost ($/h)	676.762	648.0618
P_{LOSS} (MW)	5.20730	4.2542
CPU time (s)	189.000	188.164

Table 5.2 provides the obtained optimal values for the control variables for the proposed algorithms, i.e., GWO and OKHA [1]. The convergence profile of GWO is illustrated in Figure 5.6, and the minimal active power loss as derived from GWO is 1.5710 MW, which is 1.11% smaller than the best-proposed result OKHA [1].

Analysis of the fuel cost is calculated by KHA and GWO (with the valve point effect) and the settings for the control variables.

CONCLUSIONS

This study formulates a nonlinear optimization problem with both equality and inequality constraints of the power system and proposes a Grey wolf-based optimization algorithm for tackling OPF problems combining two FACTS devices, TCSC and TCPS. Consideration is given to the two primary aim functions of minimizing gearbox loss and fuel costs. On a modified IEEE 30 bus test system, the proposed algorithm is evaluated. The suggested algorithm's simulation results demonstrate improved convergence speed and avoidance of being locked into local optimum solutions. When the results from this study are contrasted with some recent algorithms that have been discussed in the literature for OPF problems, the results show better outcomes, demonstrating the potential for this methodology to be applied to more complicated situations in the future.

REFERENCES

1. A. Abodllahi,A.A. Ghadimi, M.R. Miveh, F Mohammadi, F. Jurado "Optimal power flow incorporating FACTS devices and stochastic wind power generation using Krill Herd algorithm", *Electronics*, 9, 2020, 1043.
2. S. Sharma, B.S.A. Prabhakaran, M. Chavan, R. Srivastava "A hybrid swarm intelligence approach for resolving reactive power dispatch issues in power system: Optimal placement and sizing of UPFC", *Advances in Engineering Software*, 170, 2022, 103149.
3. K.M.K. Reddy, A.K. Rao, R.S. Rao "An improved Grey Wolf algorithm for optimal placement of unified power flow controller", *Advances in Engineering Software*, 173, 2022, 103187.
4. X. Rui, M. Sahraei Ardakani, "A successive flow direction enforcing algorithm for optimal operation of variable-impedance FACTS devices", *Electric Power Systems Research*, 211, 2022, 108171.
5. U. Guvenc, S. Duman, Hamdi, T. Kahraman, S. Aras, M.Katı "Fitness–distance balance based adaptive guided differential evolution algorithm for security-constrained optimal power flow problem incorporating renewable energy sources", *Applied Soft Computing*, 108, 2021, 107421.
6. P. Pareek, H.D. Nguyen "Non-parametric joint chance-constrained OPF via maximum mean discrepancy penalization non parametric joint chance constrained OPF via maximum mean discrepancy penalization", *Electric Power Systems Research*, 212, 2022, 108482.

7. I.I. Avramidis, G. Cheimonidis, P. Georgilakis "Ineffective control actions in OPF problems: Identification, suppression and security aspects", *Electric Power Systems Research*, 212, 2022, 108228.
8. I. Tyuryukanov, M. Popov, J.A. Bos, M.A.M.M. van der Meijden, V. Terzija "New cycle-based formulation, cost function, and heuristics for DC OPF based controlled islanding new cycle-based formulation, cost function, and heuristics for DC OPF based controlled islanding", *Electric Power Systems Research*, 212, 2022, 108588.
9. M. Sheikhan, H. Bostani "Hybrid and modified OPFs for intrusion detection systems and large-scale problems optimum-path forest", *Theory, Algorithms, and Applications*, 2022, 109–136. ISBN 9780128226889.
10. H. Bostani, M. Sheikhan, B. Mahboobi "A strong coreset algorithm to accelerate OPF as a graph-based machine learning in large-scale problems", *Information Sciences*, 555, 2021, 424–441.
11. M.A. Peñalba, T.L. Nguyen, R. Caire, A. Sumper, S.G. Arellano, Q.T. Tran "General form of consensus optimization for distributed OPF in HVAC-VSC-HVDC systems", *International Journal of Electrical Power & Energy Systems*, 121, 2020, 106049
12. S. Mirjalili, S.M. Mirjalili, A. Lewis, "Grey wolf optimizer", *Advances in Engineering Software*, 69, 2014, 46–61.
13. R.P. Singh, V. Mukherjee, S.P. Ghoshal, "Particle swarm optimization with an aging leader and challenger's algorithm for optimal power flow problem with FACTS devices", *International Journal of Electrical Power and Energy Systems*, 64, 2015, 1185–1196.
14. P.P. Biswas, P.N. Suganthan, Gehan A.J. Amaratunga "Optimal power flow solutions incorporating stochastic wind and solar power", *International Journal on Energy Conversion and Management*, 148, 2017, 1194–1207.

Chapter 6

Solving of combined heat and power economic dispatch problem using evolutionary technique considering prohibited operating zone

Sameer Sharma, Pramit Ghosh, Chandan Paul, and Sunanda Hazra

INTRODUCTION

When thermal power is generated, heat is discharged into the atmosphere by cooling towers, flue gas, or other means in every thermal power-producing plants. Thus, power generation units' energy efficiency drops to between 50% and 60%, and the atmosphere becomes contaminated. Consequently, byproducts (NOX, SOX, SO_2, and CO_2) are released during heating.

The combined heat and power economic dispatch (CHPED) problem must be understood in order to study power systems. It lowers the cost of manufacturing and the amount of pollutants released into the atmosphere by utilizing the steam's residual heat. This could increase the overall productivity of the electricity system. In a CHPED system, the recuperation of heat steam-powered apparatus uses chillers to recover heat lost during cooling and steam production. One kind of cogeneration system that simultaneously produces electricity and processes heat is called CHPED.

The realistic existence of loading at the valve point and restricted operational regions in thermal plants result in a non-linear and intermittent cost function regarding the thermal unit. For simplicity, the price-function of heat, power, and cogeneration units has been analyzed using mathematical techniques. A quadratic function describes these cost functions. Furthermore, restricted operating zones may be established for the CHPED problem's power generating units due to specific component restrictions, like the generator shaft. In light of a unit with forbidden operating zones, the entire system consists of two workable suboperating regions. The operating equality and inequality restrictions increase the complexity and discontinuity of the CHPED optimization problem system as a whole. In order to provide the best possible electric power at the lowest possible fuel cost, a quality solution must be found by utilizing a

 DOI: 10.1201/9781003628279-6

variety of optimization approaches in order to satisfy the necessary power demand and other limitations.

Many academics have published their findings on CHPED throughout the last 20 years. To solve CHPED, researchers employ a range of optimization techniques. These methods can be divided into two groups: mathematically derived sophisticated algorithms for optimization and iterative traditional mathematical optimization algorithms. Many academics have successfully applied traditional algorithms, like the lambda iteration method (LIM) [2], relaxation of Lagrangian [1], etc. A simulation of probability creation was presented by Larsen et al. approach in [3] in order to resolve the CHPED issue, which involves sampling to determine probabilities. According to Thomson et al. [4], a statistical process control approach was used to solve the CHPED problem, often known as statistical method quality control.

In general, for fuel cost curves with linear features, these methods provided the best optimal solution. These approaches, however, are not likely to resolve the practical CHPED problem, despite the fact that it has complicated and non-convex features for valve point loading with the forbidden working zone.

However, a variety of clever optimization strategies, including heuristic approaches, have been used to provide precise dispatch results for the CHPED problem. Song et al. [5] presented an enhanced ant colony optimization (ACO) technique to address the CHPED issue in the power system. Evolutionary programming (EP) was established by Wong and Algie [6] to address CHPED issues. In order to study the CHPED, Sudhakaran et al. [7] predicted using combining evolutionary algorithms for cheap tabu search and certain integral solutions dispatch. Harmony search (HS) is a strategy established in order to determine the best answer. Only units that generate power were included in this technique's application of the cubic cost function [8]; consequences of valve points were excluded. Multi-objective particle swarm optimization (MPSO) was established by Wang et al. [9] to reduce the continuous functions in the CHPED problem. Subbaraj et al. described a self-adaptive real-coded genetic algorithm (SARGA) that makes use of the simulated binary crossover (SBX) operator [10] to address the CHPED issue. Basu et al. created the bee colony optimization (BCO) method [11] to analyze CHPED while taking network losses into account at the valve location. In order to address the Economic Load Dispatch (ELD) problem, Shaw et al. created the seeker optimization algorithm (SOA) in Ref. [12]. This method is based on the idea of mimicking human uncontrolled searching behavior. Hota and associates [13] suggested a fuzzy-based modified bacterial foraging algorithm (MBFA) to assess multi-objective Economic Emission Load Dispatch (EELD) situations, hence improving the global searching ability. Simulated annealing, or SA, was created by Basu et al. [14] to solve Combined Economic Emission Load Dispatch (CEELD) difficulties. Discrete search spaces can use this strategy. Roy et al. [15] created oppositional teaching learning-based optimization (OTLBO) to investigate the CHPED problem. The optimization of spider monkey approach was proposed by Bansal et al. [16] to lower the price of producing electricity and the release of a CHPED issue. This optimization method stems from how spider monkeys forage for food. A capable combined heat and power algorithm was proposed by Rong et al. [17]. Tarafdar and associates [18] successfully established an improved group search optimization (IGSO) method for analyzing the CHPED problem in power structures. Shaw et al. first proposed the opposition-based gravitational search algorithm (OGSO) [19] for the reactive power solution transfers. Bhattacharjee et al. [20] introduced oppositional real improvement of chemical reactions by coding for the ELD. Mantegna [21] developed a novel technique using the Mantegna algorithm and levy flights. A model of spot prices was used by Elberg and colleagues [22] to account for the need for wind and electricity generation and faulty components. Javidy and associates [23] introduced an optimization method according to the ion's random motion within the atmosphere. The way this search engine behaves is similar to the allure and repulsion between both cations and anions. A novel crisscross optimization (NCSO) technique was presented by Meng et al. [24] to address the CHPED problem. Mirjalili et al. introduced the moth–flame optimization (MOF) approach [25]. The development of it came from studying moth within the atmosphere. The sine cosine algorithm (SCA), which is population-based, has been presented by Mirjalili et al. [26] where a mathematical method according to the cosine and sine functions was used to investigate the search procedure. Particle swarm optimization with unique inertia factor (PSO-UIF) was introduced by Ghorbani et al. [27] as a solution for CHPED difficulties. Jayakumar et al. introduced Grey Wolf optimization (GWO), which is a novel method for solving CHPED issues [28]. Haghrah et al. [29] used an enhanced Muhlenbein mutation in conjunction with a real-coded genetic algorithm to seek for the most effective remedy to the CHPED problem. Exchange market algorithm (EMA), put out by Ghorbani [30], was developed based on the performance of two market states: oscillated and balanced markets. Following each market state operation, an individual's fitness was assessed. Cuckoo search algorithm (CSA) was created by Thang et al. [31] to address the CHPED problem.

Long calculation times are a frequent issue for the majority of evolutionary algorithms used to solve non-linear problems, according to a review of the literature. Creating a basic method is necessary to get an efficient ideal resolution and hasten the CHPED problem's convergence time.

The aim of this discussion is to illustrate an effective method for tackling CHPED problems with the whale optimization algorithm (WOA). This WOA method is a more modern meta-heuristic technique that was put forth by Mirjalili et al. [32]. Aziz et al.'s research [33] has demonstrated the superiority of WOA over other currently used methods. The WOA mimics whales' innately cooperative nature. Because it includes two methods for exploitation and two phases of investigation, it is a dependable and gradient-free methodology.

Moreover, this algorithm provides a worldwide remedy, which is suitable for real solutions to avoiding the local optima. Another advantage of this approach is that there is not any structural change during the solution of optimization problems. This adaptable quality of the WOA algorithm motivates the present writers to use this recommended approach to examine the CHPED issues. The present algorithm is implemented on three test systems in order to demonstrate its validity. To prove its best performances from the obtained results, the method that was demonstrated is judged against classic PSO (CPSO) [34], which is used to solve the low dimensional problem. With time varying acceleration coefficient particle swarm optimization (TVAC-PSO) [34] where some modification has been added to PSO to overcome a few disadvantages of PSO and teaching learning-based optimization (TLBO) [35], which is predicated on pupils' increasing their knowledge through interactions with one another or their teachers. Lastly, group search optimization (GSO) was also contrasted [36]. *The search method of this optimization technique i.e. teaching learning based* optimization (TLBO) [35] is based on pupils' increasing their knowledge through interactions with one another or their teachers.. Lastly, GSO was also contrasted [36]. This optimization technique's search method is based on the way that animals hunt.

For the remaining part, the presentation is organized as follows. The CHPED issue formulation is presented in the "Problem formulation of CHPED" section. The "Whale optimization algorithm (WOA)" section provides a brief overview of the WOA method, while the "Encircling prey" section presents the various stages of the algorithm's application to the CHPED problem. The "Simulation result section" presents the simulation results and the convergence properties among the cost-functioning elements of the three test systems for the forbidden and without the restricted zone. The "Conclusion" section discusses the chapter's contributions and its potential future applications.

PROBLEM FORMULATION OF CHPED

The primary goal of the CHPED challenge is to recover waste heat from steam in an effort to reduce the price of fuel used in power generation. It must meet all requirements when producing heat and electricity. Furthermore, during power generation, the cogeneration unit's viable operation zone of heat power is also satisfied. The CHPED problem formulation takes forbidden zone and valve point effects into account for a practical scenario to verify the overall performances of CHPED. The following illustration displays the analytical version of the CHPED problem.

The CHPED problem's objective function is provided by:

$$\text{Min COST} = \sum_{i=1}^{N_p} C_{pi}(P_{pi}) + \sum_{j=1}^{N_c} C_{cj}(P_{cj}, H_{cj}) + \sum_{k=1}^{N_h} C_{hK}(H_{hK}) \tag{6.1}$$

The cost of overall generation is represented by C in the equation above. It also denotes the fuel cost function of the ith power-only unit, the production cost of cogeneration units, and the cost of heat units. The numbers indicate the number of power units, cogeneration units, and heat-only units.

The quadratic cost function is used to represent the thermal unit.

$$C_{pi}(P_{pi}) = x_{pi}(P_{pi})^2 + y_{pi}P_{pi} + z_{pi} \tag{6.2}$$

where x_{pi}, y_{pi}, and z_{pi} are the coefficient of cost function of the ith thermal unit.

$$C_{pi}(P_{pi}) = x_{pi}(P_{pi})^2 + y_{pi}P_{pi} + x_{pi} + \left|g_{pi}\ Sin\left(h_{pi} \times \left(P_{pi}^{\min} - P_{pi}\right)\right|\right. \tag{6.3}$$

Equation (6.3) shows valve point effects while accounting for the cost function equation (6.2). The valve point loading is described by a sinusoidal equation. The problem is no longer convex and non-differentiable as equation (6.3) sums the quadratic cost function with a sinusoidal factor.

where g_{pi} and h_{pi} are the ith unit's coefficients of the cost function for simulating the impacts of valve point loading.

$$C_{cj}(p_{cj}, H_{cj}) = x_{cj}(p_{cj})^2 + y_{cj}p_{cj} + z_{cj} + g_{cj}(H_{cj})^2 + h_{cj}H_{cj} + K_{cj}H_{cj}\ p_{cj} \tag{6.4}$$

$$C_{hK}(H_{hK}) = x_{hK}(H_{hK})^2 + y_{hK}H_{hK} + z_{hK} \tag{6.5}$$

where $C_{cj}(P_{cj}, H_{cj})$ depicts the cost function of the jth cogeneration unit; $C_{hK}(H_{hK})$ is the cost function of the kth heat-only unit.

The following constraints of CHPED are given below:

i. Power balance constraint

$$\sum_{i=1}^{Np} p_{Pi} + \sum_{j=1}^{N_c} p_{cj} = p_D + p_L \tag{6.6}$$

$$p_L = \sum_{i=1}^{Np}\sum_{m=1}^{Np} p_i B_{im} p_m + \sum_{i=1}^{Np}\sum_{j=1}^{N_c} p_i\ B_{ij} p_j + \sum_{j=1}^{N_c}\sum_{r=1}^{N_c} p_j B_{jr} p_r \tag{6.7}$$

$$\sum_{j=1}^{N_c} H_j + \sum_{k=1}^{N_h} H_{hK} = H_D \tag{6.8}$$

Equation (6.6) represents the balance of heat production and demand, equation (6.7) represents the transmission line's power loss, and equation (6.8) represents the balance of power production and demand.

where H_D is the thermal demand and B_{im}, B_{ij}, and B_{jr} are the loss coefficients of the transmission system.

ii. Capacity constraints

Every power generating unit, cogeneration unit, and heat unit's total output of electricity and heat should fall within the following limitation range for stable operation.

$$p_{Pi}^{\min} \le p_{Pi} \le p_{Pi}^{\max} \tag{6.9}$$

$$p_{cj}^{\min}(H_{cj}) \le p_{cj} \le p_{cj}^{\max}(H_{cj}) \quad \text{where, } j = 1,2,3,\dots,N_c \tag{6.10}$$

$$H_{cj}^{\min}(p_{cj}) \le H_{cj} \le H_{cj}^{\max}(p_{cj}) \quad \text{where, } j = 1,2,3,\dots,N_c \tag{6.11}$$

$$H_{hK}^{\min} \le H_{hK} \le H_{hK}^{\max} \quad \text{where, } K = 1,2,3,\dots,N_h \tag{6.12}$$

For power-only unit, min and max power are represented by $p_{Pi}^{\min}$ and $p_{Pi}^{\max}$, where min and max power limit of the jth cogeneration unit are defined by 2 and $p_{cj}^{\max}(H_{cj})$. Again $H_{cj}^{\min}(p_{cj})$ and $H_{cj}^{\max}(p_{cj})$ are min and max heat limit of the jth cogeneration unit, respectively, and $H_{hK}^{\min}$ and $H_{hK}^{\max}$ are min and max heat limit of the Kth heat unit, respectively.

iii. Prohibited operating zones

iv. A zone where the production of electricity is entirely forbidden is known as a banned operating zone for thermal units. When there is a chance that the oscillation of the rotor shaft and the power frequency produced during power generation will coincide, resonance occurs.

The rotor shaft may vibrate more forcefully as a result of this resonance, which might damage the shaft or make the system unstable. It is very difficult to establish the $I/P-O/P$ characteristics in the restricted zone due to the discontinuity. During actual operation, the zones should be sorted out and ignored in order to optimize the economy. This is how the unit will be explained: prohibited operating zone (POZ) operates in an impractical area.

$$p_{Pi}^{\min} \le p_{Pi} \le p_{Pi,1}^{l} \quad \text{where, } i = 1,2,3,\dots,N \tag{6.13}$$

$$p_{Pi,j-1}^{u} \le p_{Pi} \le p_{Pi,j}^{l} \quad \text{where, } j = 1,2,3,\dots,N \tag{6.14}$$

$$p_{Pi,N}^{u} \le p_{Pi} \le p_{Pi}^{\max} \quad \text{where, } i = 1,2,3,\dots,N \tag{6.15}$$

The number of forbidden zones in the relationship above is stated in terms of N for the ith unit; $p_{Pi,j-1}^{u}$ shows the creation of an upper limit; and $p_{Pi,j}^{l}$ is the forbidden zone's lower generation limit for j and $j-1$, respectively, of the ith unit.

v. Heat-power range of cogeneration units that is feasible.

For three test systems, the cogeneration units' practicable heat-power zones are as follows:

Test system 1:

$$\begin{aligned}
&1.781914894 \times H_5 - p_5 - 105.7446809 \leq 0 \\
&0.1777777784 \times H_5 + p_5 - 247.0 \leq 0 \\
&-0.169847328 \times H_5 - p_5 + 98.8 \leq 0 \\
&1.158415842 \times H_6 - p_6 - 46.88118818 \leq 0 \\
&0.151162791 \times H_6 + p_6 - 130.6976744 \leq 0 \\
&-0.067681895 \times H_6 - p_6 + 45.07614213 \leq 0
\end{aligned} \tag{6.16}$$

Test system 2:

$$\begin{aligned}
&1.781914894 \times H_{14} - p_{14} - 105.7446809 \leq 0 \\
&0.1777777784 \times H_{14} + p_{14} - 247.0 \leq 0 \\
&-0.169847328 \times H_{14} - p_{14} + 98.8 \leq 0 \\
&1.158415842 \times H_{15} - p_{15} - 46.88118818 \leq 0 \\
&0.151162791 \times H_{15} + p_{15} - 130.6976744 \leq 0 \\
&-0.067681895 \times H_{15} - p_{15} + 45.07614213 \leq 0 \\
&1.781914894 \times H_{16} - p_{16} - 105.7446809 \leq 0 \\
&0.1777777784 \times H_{16} + p_{16} - 247.0 \leq 0 \\
&-0.169847328 \times H_{16} - p_{16} + 98.8 \leq 0 \\
&1.158415842 \times H_{17} - p_{17} - 46.88118818 \leq 0 \\
&0.151162791 \times H_{17} + p_{17} - 130.6976744 \leq 0 \\
&-0.067681895 \times H_{17} - p_{17} + 45.07614213 \leq 0 \\
&0.25 \times H_{18} - p_{18} + 20 \leq 0 \\
&0.272727272 \times H_{18} + p_{18} - 60 \leq 0 \\
&2.333333333 \times H_{18} - p_{18} - 83.333333333 \leq 0 \\
&2.2 \times H_{19} - p_{19} - 9 \leq 0 \\
&0.6 \times H_{19} + p_{19} - 105 \leq 0
\end{aligned} \tag{6.17}$$

WHALE OPTIMIZATION ALGORITHM (WOA)

A new optimization method called WOA is presented by Mirjalili and Lewis and is documented in the study by Mirjalili et al. [32]. The WOA was created using a unique humpback whale hunting method. This hunting approach can be categorized into several forms, such as surrounding the prey, using a bubble net to hunt, and searching for prey during the exploration phase. These forms are thought to represent distinct stages in the WOA algorithm. The various methods of hunting are examined one by one as follows.

ENCIRCLING PREY

According to the WOA algorithm, the target prey is the best option for encircling the prey. Whales that are humpbacks are looking for prey that will provide the optimum answer. One by one, they encircle the prey's location. When a search agent locates the intended prey, other search agents often adjust their current position to align with the ideal location.

$$\begin{aligned} &\vec{X}(T+1) = \vec{X}^*(T) - \vec{K}.\vec{L} \\ &\vec{L} = \left| \vec{M} \cdot \vec{X}*(T) - \vec{X}(T) \right| \end{aligned} \tag{6.18}$$

In the above equation, $\vec{K}$ and $\vec{M}$ are coefficient vectors and are represented by:

$$\vec{K} = 2 \cdot \vec{a} \cdot \vec{r} - \vec{a} \tag{6.19}$$

$$\vec{M} = 2 \cdot \vec{r} \tag{6.20}$$

where current iteration is indicated by T; the position vector and the position of the optimum solution are defined by $\vec{X}(T)$ and $\vec{X}*(T)$, respectively; $\vec{L}$ signifies the gap among the location of $\vec{X}*(T)$ and $\vec{X}(T)$; and $\vec{K}$ and $\vec{M}$ are coefficient vectors. The random value of $\vec{K}$ is in the range within $[-a, a]$ and its value is setting down in the range $[-1, 1]$; the value of $\vec{K}$ and $\vec{a}$ are linearly decreased from 2 to 0; and $\vec{r}$ is a random vector and the value of $\vec{r}$ is within the range $[0, 1]$.

BUBBLE NET HUNTING METHOD

The two methods of bubble net hunting are as follows: first, in shrinking encircling, humpback whales look for prey, swim around it, and create a distinct bubble that travels in a circle.

In behavior of shrinking, encircling prey $(\vec{a})$ is represented by

$$\vec{a} = 2 - t\frac{2}{\text{Maxiteration}} \tag{6.21}$$

In equation (6.21), t denotes the number of iterations. Maxiteration signifies the max number of iterations. $\vec{K}$ is also decreased by $\vec{a}$. The updated location of $\vec{K}$ is acquired between actual location and present best location of the agent.

One updates their position in a spiral pattern among the locations of the whales and prey in order to track the helix-shaped movement of humpback whales. The whales' most recent position is shown below.

$$\vec{X}(T+1) = \vec{L}\cdot e^{bn}\cos(2\pi n) + \vec{X}^*(T) \tag{6.22}$$

From the above equation, $\vec{L} = \left|\vec{X}^*(T) - X(T)\right|$ denotes the distance of the ith whale to fittest solution prey; the constant b and the random integer n in [−1, 1] dictate the shape of the logarithmic spiral.

When whales are positioning themselves to maximize their catch, they swim around their prey. The whales' coordinated activity is expected to occur 50% of the time between the spiral model and the declining encircling mechanism in order to update the whales' location.

$$\vec{X}(T+1) = \text{Shrinking encircling, if } P < 0.5$$

$$\vec{X}(T+1) = \text{Spiral encircling, if } P > 0.5$$

In above equation, P is a random number [0, 1], during the optimization process.

SEARCH FOR PREY (EXPLORATION PHASE)

Rather than updating the search agent's location, a random number is chosen and searched. Therefore, random numbers larger than 1 or smaller than −1 can be used with the vector. The following equation can be used to explain this search process.

$$\vec{L} = \left|\vec{C} \cdot \vec{X}_{\text{rand}} - \vec{X}\right|$$
$$\vec{X}(T+1) = \vec{X}_{\text{rand}} - \vec{K}\vec{L} \tag{6.23}$$

QUASI WHALE OPTIMIZATION ALGORITHM (QWOA) STEPS FOR CHPED

Several steps of the WOA that are used to solve the CHPED problem are as follows:

Step 1: Set the cogeneration units' active power and power completely at random. Reset the heat-only units and the cogeneration heat to a random initial value. They should be operating within their maximum and lowest initialized values. Give the WOA technique's input parameters as well.

Step 2: It is determined whether the cogeneration units can provide enough power and heat. The cogeneration units' properties of heat versus power boundary criterion are satisfied by this feasibility. In the event that the outcome is not viable, a new feasible outcome is generated to replace the unfeasible one.

Step 3: Ignoring power generation in the forbidden operating zone yields the best secure approach for practical operation. The left and right banned subzones are the two subzones that make up each forbidden operating zone. The thermal generating unit can now function in the forbidden zone to the left or right. The generating unit moves toward the right subzone's higher limit if it is operating in the right subzone. Once more, it moves toward the bottom boundary of the left subzone when it operates in that area.

Step 4: Assess the heat and power generation of the slack heat and power units, respectively, and determine if these values fall within the range of the maximum and minimum working limits. Additionally, it is verified if the slack unit's power generation satisfies the restrictions on the restricted operating zone; if not, the slack bus solution is dismissed. The set of infeasible options is replaced with the newly generated feasible solution.

Step 5: Each feasible solution set's fitness value must be computed. After that, arrange the fitness values in ascending order among all of the population that was created.

Step 6: A select few of these solutions are regarded as elite options. Since these fixes will yield the best outcome, the issue must be avoided.

Step 7: To change the independent variables, the various WOA hunting techniques are once more applied to non-elite solutions. The optimal position is determined by the whales' pursuit of their prey, which is the target of the WOA algorithm. The ultimate solution can be found at the target prey's location. Using the previously mentioned methods, the other search agents update their location by following the location of the top search agent.

Step 8: Determine the CHPED problem's limiting value for the independent variables. Operating constraints should apply to this limiting value. If the independent variable exceeds its maximum value, it is set at maximum, and if it falls below the minimum value of the limiting range, it is set as minimum.

Step 9: Determine whether slack units are feasible. If the problem is not resolved, the best workable solution that is currently generated takes its place. Duplicate solutions are shifted to make room for newly created solutions.

Step 10: Proceed to the ultimate optimal solution if the condition for ending the iteration is met.

Step 11: If not, proceed to step 7.

SIMULATION RESULT

The current CHPED simulation study uses three test systems. The WOA algorithm is used on these CHPED test devices to demonstrate its efficacy and efficiency. By contrasting the outcomes of the WOA algorithm with those of CPSO [34], TVAC-PSO [34], TLBO [35], and GSO [36], it has been demonstrated how superior the provided method is. MATLAB 7.8 is used to execute the simulation and test it. The machine, which has a newer Core i5 processor with 2.5 GHz internal memory and 4 GB RAM, is utilized to run MATLAB. The following presents the simulation results and computational time for test systems 1, 2, and 3 using the suggested algorithm. Additionally, it covered how various cogeneration units generate heat and power within a realistic operating range. The current WOA method yields the greatest results with the least amount of computational time for population sizes of 50. For every scenario with and without a forbidden zone, each population is run 100 times.

TEST SYSTEM 1

Seven units make up this system: two combined heat and power (CHP) units, one heat-only unit, and four power-only units. The heat and power generation of both cogeneration units for this system are displayed in Figures 6.1 and 6.2, and they fall within the system's practical working parameters. In the beginning, the problem's solution assumes a 600-MW power demand and a 150-MWth heat requirement. Figure 6.3 shows the convergence graph of the proposed WOA algorithm for the forbidden zone. The simulation runs 100 times, and Tables 6.1 and 6.2 explain the mean of the answer, which is the average of the 100 simulation outcomes. Table 6.1

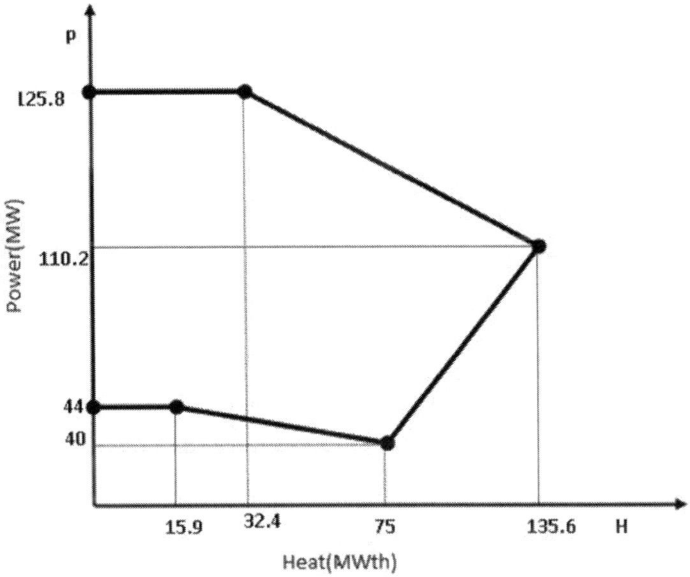

Figure 6.1 Feasible operation region of cogeneration units (6th unit for case I and 15th and 17th units for case II).

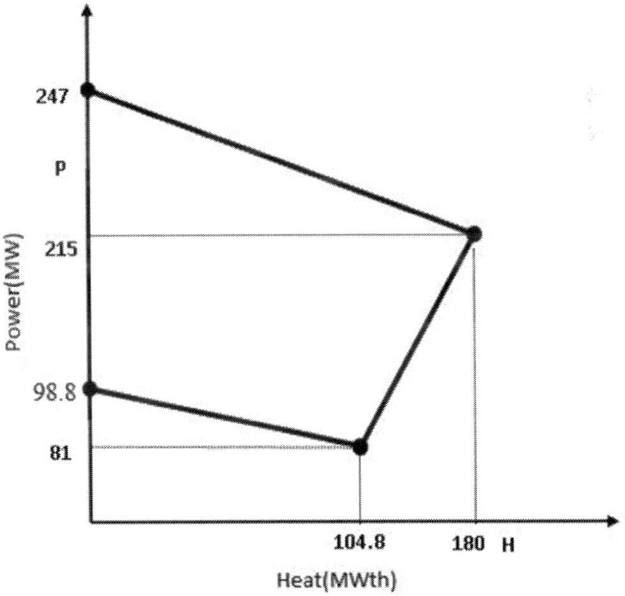

Figure 6.2 Feasible operation region of cogeneration units (5th unit for case I and 14th and 16th units for case II).

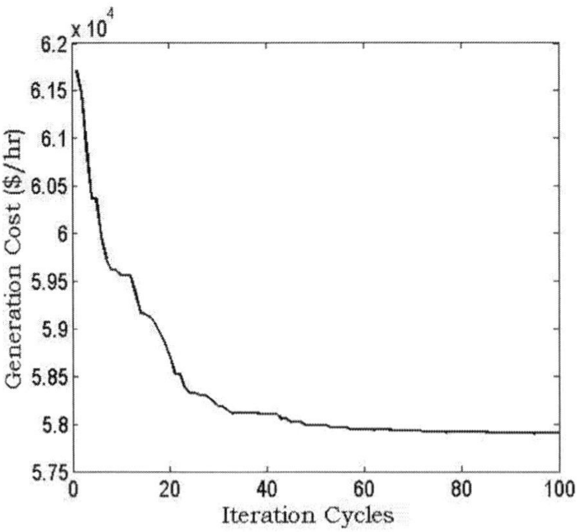

Figure 6.3 Cost convergence profile of quasi whale optimization algorithm (QWOA) for test system I with prohibited operating zone.

(excluding the forbidden zone) illustrates that the lowest cost per hour using WOA is 10094.2091 $/hr. This demonstrates unequivocally that WOA results in far lower costs than do the other methods. Furthermore, WOA takes 2.3216 seconds to compute, which indicates that the computing time for the suggested WOA strategy is significantly less than that of the other strategies listed in Table 6.1. It is evident from Table 6.2 (which includes the forbidden zone) that the WOA algorithm outperforms other algorithms in terms of economy and computing speed.

The results of WOA are compared with the results utilizing GSO in order to determine whether the presented WOA technique is superior to previous optimization techniques [36]. Convergence graph in Figure 6.3 shows the computed generation costs after 100 trial runs. When the answer, after 60 iterations, comes within 0.02% of the best solution cost of the forbidden zone, these outcomes demonstrated the stability of the solutions provided by the WOA approach. It follows that the global optimal solution was obtained using the WOA technique that was given.

TEST SYSTEM 2

There are a total of 24 units in the test system that is being displayed. These 24 units are composed of 6 CHP units, 5 heat-only units, and 13 power-only units. The system is examined using the WOA optimization method. The initial power demand for this test is 2350 MW, while the heat demand is

Table 6.1 MATLAB results achieved by different approaches of test system 1 (without prohibited zone)

Control variables (MW and MWth)	*CPSO [34]*	*TVAC-PSO [34]*	*TLBO [35]*	*GSO [36]*	*WOA*	*QWOA*
Tp_1	75.0000	47.3383	45.2660	45.6188	45.6072	45.5834
Tp_2	112.3800	98.5398	98.5479	98.5401	98.5398	98.5501
Tp_3	30.0000	112.6735	112.6786	112.6727	112.6735	112.3468
Tp_4	250.0000	209.81582	209.8284	209.8154	209.8158	208.7781
Tp_5	93.2701	92.3718	94.4121	94.1027	94.1021	94.0886
Tp_6	40.1585	40.0000	40.0062	40.0001	40.0001	39.9867
h_5	32.5655	37.8467	25.8365	27.6600	27.6596	27.5643
h_6	72.6738	74.9999	74.9970	74.9987	75.0000	74.8967
h_7	44.7606	37.1532	49.1666	47.3413	47.3404	46.3324
		Statistical analysis				
Minimum cost ($/hour)	10325.33	10100.32	10094.83	10094.2318	10094.2091	
Mean cost ($/hour)	–	–	10114.15	10095.6615	10094.8214	
Maximum cost ($/hour)	–	–	10133.61	10097.2406	10095.9102	
Computational time (seconds)	–	–	2.86	2.4203	2.3216	

Table 6.2 Simulation results achieved by various approaches of test system 1 (with prohibited zone)

Control variables (MW and MWth)	*GSO [36]*	*WOA*	*QWOA*
Tp_1	44.1443	44.3065	44.4023
Tp_2	100.0023	98.5373	99.5534
Tp_3	112.6752	112.6795	112.6623
Tp_4	209.8148	209.8155	209.7845
Tp_5	94.1126	95.3859	93.2312
Tp_6	40.0004	40.0137	40.9867
h_5	27.6002	20.1011	20.2345
h_6	74.9975	75.0117	74.2326
h_7	47.4023	54.8872	53.9967
Cost ($/hour)			
	Statistical analysis		
Minimum cost ($/hour)	10101.3483	10098.4554	10097.3423
Mean cost ($/hour)	10102.2168	10099.8627	10099.4532
Maximum cost ($/hour)	10103.7203	10100.7893	10100.1278
Computational time (seconds)	2.5903	2.3715	2.3512

1250 MWth. Figures 6.4 and 6.5 depict the convergence profile of the cost function from the suggested WOA algorithms for the systems with and without restricted zones. Figures 6.1, 6.2, 6.6, and 6.7 depict the heat and electricity generation of the cogeneration units in practical operating regions.

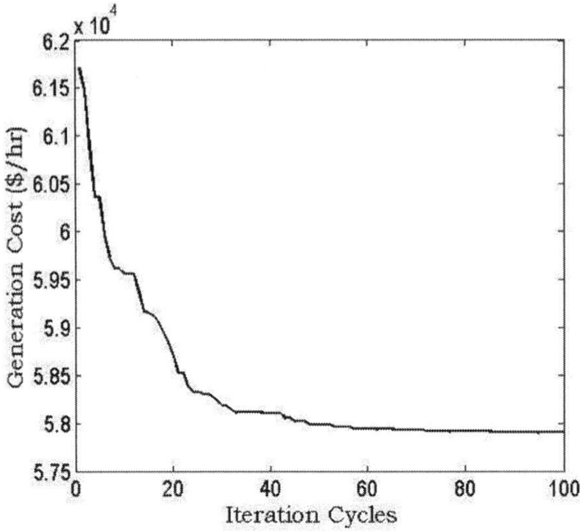

Figure 6.4 Cost convergence profile of QWOA for test system 2 without prohibited operating zone.

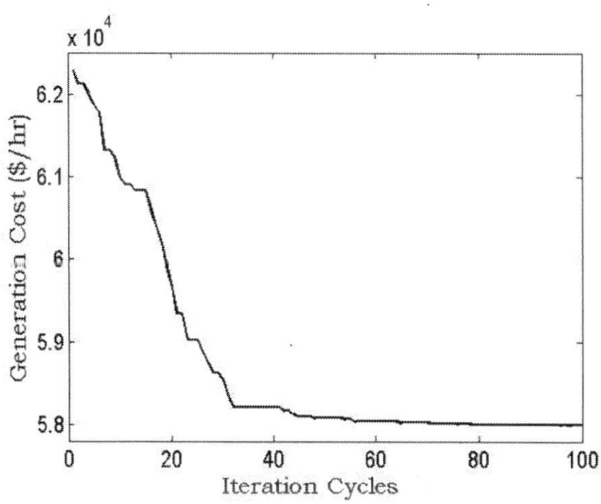

Figure 6.5 Cost convergence profile of QWOA for test system 2 with prohibited operating zone.

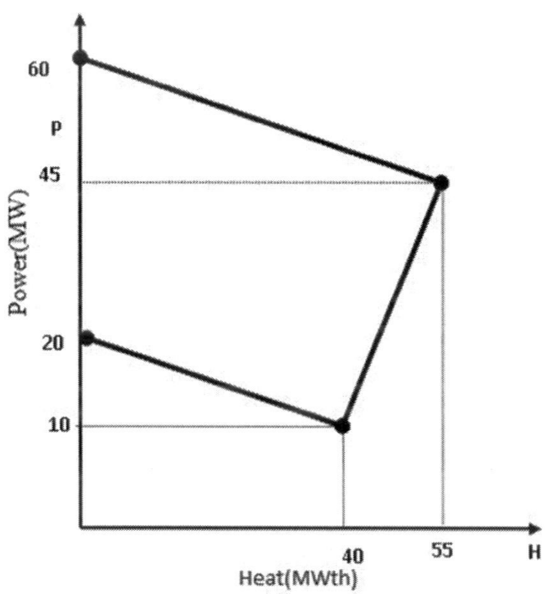

Figure 6.6 Feasible operation region of cogeneration units (18th unit for case II).

Table 6.3 displays the mean, minimum, maximum, and computation time from the convergence graph for 100 iterations of the presented approach and alternative algorithms, such as CPSO [34], TVAC-PSO [34], TLBO [35], and GSO [36], for the system without a restricted zone. Table 6.4 displays the mean, minimum, maximum, and calculation time from the

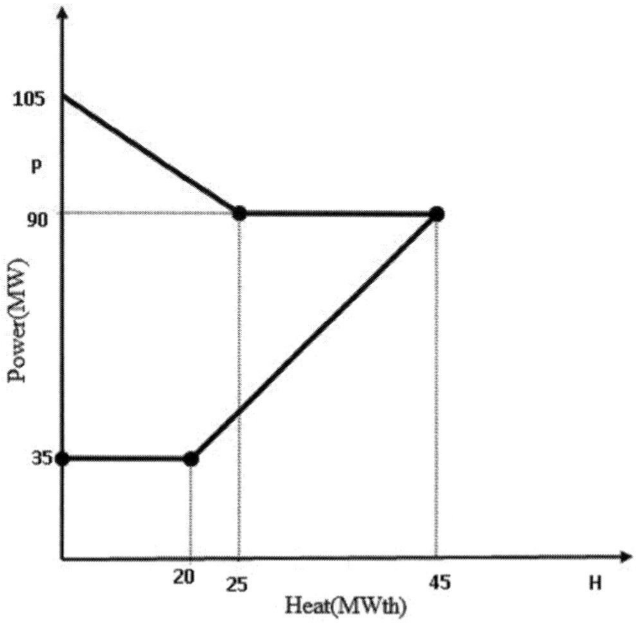

Figure 6.7 Feasible operation region of cogeneration units (19th unit for case II).

convergence graph for 100 iterations of the given algorithm and with GSO [36] for the system with the restricted zone. Tables 6.3 and 6.4 explain various algorithmic outcomes, which demonstrate how much better the suggested WOA performs in the field of cost minimization than alternative optimization strategies. The rates are 57997.0697 $/hr for the forbidden zone and 57898.6023 $/hr for the non-prohibited zone.

It is calculated as follows: 57997.0697 $/hr for the banned zone and 57898.6023 $/hr for the non-prohibited zone. Following that, the computation time is 5.2865 seconds for the banned zone and 5.3214 seconds for the non-prohibited zone.

CONCLUSION

This talk describes the successful implementation of a novel, effective, evolutionary-based whale optimization technique to tackle the CHPED problem. To assess the viability and effectiveness of the suggested optimization strategy, three test systems are subjected to a mathematical simulation utilizing the QWOA algorithm. Other established methodologies reported in this chapter are compared with the output results achieved by the reported QWOA technique for these systems. The QWOA method reduces both the

Table 6.3 MATLAB results achieved by various approaches of test system 2 (without prohibited zone)

Control variables (MW and MWth)	Algorithms						Control variables (MW and MWth)	Algorithms					
	CPSO [34]	TVAC-PSO [34]	TLBO [35]	GSO [36]	WOA	QWOA		CPSO [34]	TVAC-PSO[34]	TLBO [35]	GSO [36]	WOA	QWOA
Tp_1	680.0000	538.5587	628.3240	538.2192	538.5587	537.5678	Tp_{16}	117.4854	88.3514	84.771	81.262	81.4314	82.3423
Tp_2	0.0000	224.4608	227.3588	298.7686	299.2377	299.7721	Tp_{17}	45.9281	40.5611	40.5874	40.0119	40.651	40.6412
Tp_3	0.0000	224.4608	225.9347	298.9086	224.9947	224.7988	Tp_{18}	10.0013	10.0245	10.001	10.0011	10	10.0001
Tp_4	180.0000	109.8666	110.3721	110.1919	109.8687	108.8546	Tp_{19}	42.1109	40.4288	31.0978	35.0012	30	30
Tp_5	180.0000	109.8666	110.2461	110.0846	109.8684	109.8765	h_{14}	125.2754	108.9256	105.6717	105.211	105.4388	104.4388
Tp_6	180.0000	109.8666	160.1761	110.1390	109.8684	108.8323	h_{15}	80.1175	75.4844	76.2843	76.5306	78.5139	77.5239
Tp_7	180.0000	109.8666	108.3552	110.1045	109.8741	109.1487	h_{16}	125.2754	108.9256	106.9125	105.5119	105.0421	106.0421
Tp_8	180.0000	109.8666	110.5379	110.2444	109.8724	109.7824	h_{17}	80.1174	75.484	75.5061	75.4706	75.562	75.6521
TP_9	180.0000	109.8666	110.5672	110.1992	109.8767	109.6785	h_{18}	40.0005	40.0104	39.9986	39.9999	40	40
Tp_{10}	50.5304	77.5210	75.7562	77.4989	77.4495	78.9543	h_{19}	23.2322	22.4676	18.2205	18.4014	17.7273	17.2775
Tp_{11}	50.5304	77.5210	41.8698	77.7367	77.4027	77.2723	h_{20}	415.9815	458.702	468.2278	468.9029	467.7172	468.6268
Tp_{12}	55.0000	120.0000	92.4789	55.1036	92.4311	92.1134	h_{21}	60	60	59.9867	59.9995	60	60
Tp_{13}	55.0000	120.0000	57.5140	55.1107	92.4056	92.5642	h_{22}						
Tp_{14}	117.4854	88.3514	82.5628	81.0624	82.1383	82.8414	h_{23}						
Tp_{15}	45.9281	40.5611	41.4891	40.3515	44.0706	44.1214	h_{24}						

(*Continued*)

Table 6.3 (Continued) MATLAB results achieved by various approaches of test system 2 (without prohibited zone)

Control variables (MW and MWth)	*Algorithms*						*Control variables (MW and MWth)*	*Algorithms*					
	CPSO [34]	*TVAC-PSO [34]*	*TLBO [35]*	*GSO [36]*	*WOA*	*QWOA*		*CPSO [34]*	*TVAC-PSO[34]*	*TLBO [35]*	*GSO [36]*	*WOA*	*QWOA*

	Statistical analysis					
	CPSO [34]	TVAC-PSO [34]	TLBO[35]	GSO[36]	WOA	QWOA
Minimum cost ($/hr)	59736.2635	58122.7460	58006.9992	57843.5191	57898.6023	57896.5432
Mean cost ($/hr)	59853.4780	58198.3106	57849.3017	57900.2137	57888.4123	57888.4123
Maximum cost ($/hr)	60076.6903	58359.5520	57857.7938	57903.4420	57892.3252	57892.3252
Computational time (Sec)	8.00	7.84	5.4106	5.2865	5.1412	5.1412

Table 6.4 Simulation results achieved by various approaches of test system 2 (with prohibited zone)

Control variables (MW and MWth)	*Algorithms*			*Control variables (MW and MWth)*	*Algorithms*		
	GSO [36]	*WOA*	*QWOA*		*GSO [36]*	*WOA*	*QWOA*
Tp_1	628.3274	628.3185	628.3185	Tp_{16}	86.1718	81.1391	80.1931
Tp_2	299.2273	224.0183	223.0233	Tp_{17}	44.9263	44.1953	43.1593
Tp_3	0.0015	149.4003	148.4002	Tp_{18}	10.0039	10.0016	10.0015
Tp_4	60	159.6179	159.5678	Tp_{19}	36.4003	32.2199	33.2192
Tp_5	159.7333	159.7363	159.3763	h_{14}	113.2357	104.8069	103.7069
Tp_6	159.7334	109.4761	109.7461	h_{15}	83.9529	78.4535	77.2335
Tp_7	159.7373	60.0157	59.0127	h_{16}	107.7004	104.8775	103.7975
Tp_8	60.0025	109.5923	108.5893	h_{17}	79.2794	78.6214	79.2614
Tp_9	159.7361	159.3956	159.3956	h_{18}	40.0006	39.9996	40.8396
Tp_{10}	77.4139	77.2	78.2	h_{19}	20.6318	18.7361	19.3716
Tp_{11}	114.8064	40.0084	40.0083	h_{20}	445.2006	465.0446	464.0245
Tp_{12}	55.0003	91.1838	90.8134	h_{21}	60.0000	59.9996	60.8396
Tp_{13}	92.4065	89.4657	89.6475	h_{22}	59.9988	59.9975	59.9957
Tp_{14}	96.0351	81.0138	80.0318	h_{23}	120.0000	119.7252	120.2752
Tp_{15}	50.3365	44.0014	43.0104	h_{24}	119.9999	119.7381	119.3792

Statistical analysis

	GSO [36]	*WOA*	*QWOA*
Minimum cost ($/hour)	58110.0900	57997.0697	57995.1278
Mean cost ($/hour)	58114.6060	57999.3115	57987.3214
Maximum cost ($/hour)	58119.1635	58003.5016	57985.4532
Computational time (seconds)	5.8017	5.3214	4.0012

generating cost and the computational time; this is demonstrated by the technique's simulation results. Thus, in the future, these QWOA features will offer the greatest answer for a workable base power system to handle optimization difficulties.

REFERENCES

1. G. Tao, M.I. Henwood, O.M. Van. "An algorithm for heat and power dispatch," *IEEE Trans Power Syst*; Vol. 11, No. 4, pp. 1778–1784, 1996.
2. C.L. Chen, S.C. Wang. "Branch and bound scheduling for thermal generating units," *IEEE Trans Energy Convers*; Vol. 8, No. 2, pp. 184–189, 1993.
3. H.V. Larsen, H. Palsson, H.F. Ravn. "Probabilistic production simulation including combined heat and power plants," *Electric Power Syst*; Vol. 48, No. 1, pp. 45–56, 1998.
4. M. Thomson, P.M. Twigg, B.A. Majeed, N. Ruck. "Statistical process control based fault detection of chp units," *Control Eng Practice*; Vol. 8, No. 1, pp. 13–20, 2000.
5. Y.H. Song, C.S. Chou, T.J. Stonham. "Combined heat and power economic dispatch by improved ant colony search algorithm," *Electr Power Syst Res*; Vol. 52, No. 2, pp. 115–21, 1999.
6. K. P. Wong, C. Algie. "Evolutionary programming approach for combined heat and power dispatch," *Electric Power Syst Res*; Vol. 61, pp. 227–232, 2002.
7. M. Sudhakaran, S.R. Slochanal. "Integrating genetic algorithms and tabu economic dispatch search for combined heat and power," *Power Syst Stab Control*; pp. 67–71, 2003.
8. A. Vasebi, M. Fesanghary, S. Bathaee. "Combined heat and power economic dispatch by harmony search algorithm," *Electric Power Energy System*; Vol. 29, pp. 713–719, 2007.
9. L.F. Wang, C. Singh. "Stochastic combined heat and power dispatch based on multi-objective particle swarm optimization," *Int J Electr Power Energy Syst*; Vol. 30, pp. 226–34, 2008.
10. P. Subbaraj, R. Rengaraj, S. Salivahanan. "Enhancement of combined heat and power economic dispatch using self-adaptive real-coded genetic algorithm," *Appl Energy*; Vol. 86, No. 6, pp. 915–21, 2009.
11. M. Basu. "Bee colony optimization for combined heat and power economic dispatch," *Expert Syst Appl*; Vol. 38, pp. 13527–13531, 2011.
12. B. Shaw, V. Mukherjee, S.P. Ghoshal. "Solution of economic dispatch problems by seeker optimization algorithm," *Expert Syst Appl*; Vol. 39, No. 1, pp. 508–519, 2012.
13. P.K. Hota, A.K. Barisal, R. Chakrabarti. "Economic emission load dispatch through fuzzy based bacterial foraging algorithm," *Int J Electr Power Energy Syst*; Vol. 32, pp. 794–803, 2010.
14. M. Basu. "A simulated annealing-based goal attainment method for economic emission load dispatch of fixed head hydrothermal power systems," *Electr Power Energy Syst*; Vol. 27, pp. 147–53, 2005.
15. P.K. Roy, C. Paul, S. Sultana. "Oppositional teaching learning based optimization approach for combined heat and power dispatch," *Electr Power Energy Syst*; Vol. 57, pp. 392–403, 2014.

16. J.C. Bansal, H. Sharma, S.S. Jadon, M. Clerc. "Spider monkey optimization algorithm for numerical optimization," *Memetic Comput*; Vol. 6, pp. 31–47, 2014.
17. A. Rong, J.R. Figueira, R. Lahdelma. "An efficient algorithm for bi-objective combined heat and power production planning under the emission trading scheme," *Energy Convers Manag*; Vol. 88, pp. 525–534, 2014.
18. M. Hagh, Tarafdar, S. Teimourzadeh, M. Alipour, P. Aliasghary. "Improved group search optimization method for solving CHPED in large scale power systems," *Energy Convers Manag*; Vol.80, pp. 446–56, 2014.
19. B. Shaw, V. Mukherjee, S.P. Ghoshal. "Solution of reactive power dispatch of power systems by an opposition-based gravitational search algorithm," *Electr Power Energy Syst*; Vol. 55, pp. 29–40, 2014.
20. K. Bhattacharjee, A. Bhattacharya, S. Dey Halder nee. "Oppositional real coded chemical reaction optimization for different economic dispatch problems," *Electr Power Energy Syst*; Vol. 55, pp. 378–91, 2014.
21. R.N. Mantegna. "Fast accurate algorithm for numerical simulation of Levy stable stochastic processes," *Phys Rev E*; Vol. 49, pp. 4677–83, 1994.
22. C. Elberg, S. Hagspiel. "Spatial dependencies of wind power and interrelations with spot price dynamics," *Eur J Oper Res*; Vol. 241, No. 1, pp. 260–272, 2015.
23. B. Javidy, A. Hatamlou, S. Mirjalili. "Ions motion algorithm for solving optimization problem," *Appl Soft Comput*; Vol. 32, pp.72–79, 2015.
24. A. Meng, P. Mei, H. Yin, X. Peng, Z. Guo. "Crisscross optimization algorithm for solving combined heat and power economic dispatch problem," *Energy Convers Manag*; Vol. 105, pp. 1303–1317, 2015.
25. S. Mirjalili. "Moth-flame optimization algorithm: A novel nature-inspired heuristic paradigm," *Knowl-Based Syst*; Vol. 89, pp. 228–249, 2015.
26. S. Mirjalili. "SCA: A sine cosine algorithm for solving optimization problems," *Knowl-Based Syst*; Vol. 96, pp. 120–133, 2016.
27. N. Ghorbani, P. Farhadi. "Combined heat and power economic dispatch problem solution using particle swarm optimization with unique inertia factor," *Majlesi J. Energy Manag*, Vol. 5, pp. 19–26, 2016.
28. N. Jayakumar, S. Subramanian, S. Ganesan, E.B. Elanchezhian."Grey wolf optimization for combined heat and power dispatch with cogeneration systems," *Electr Power Energy Syst*; Vol. 74, pp. 252–64, 2016.
29. A. Haghrah, M. Nazari-Heris, B. Mohammadi-ivatloo. "Solving combined heat and power economic dispatch problem using real coded genetic algorithm with improved Mühlenbein mutation," *Appl Therm Eng*; Vol. 99, pp. 465–475, 2016.
30. N. Ghorbani. "Combined heat and power economic dispatch using exchange market algorithm," *Int J Electr Power Energy Syst*; Vol. 82, pp. 58–66, 2016.
31. T.T. Nguyen, N.V. Dieu, B.H Dinh. "Cuckoo Search algorithm for combined heat and power economic dispatch," *Electr Power Energy Syst*; Vol. 81, pp. 204–214, 2016.
32. S. Mirjalili, A. Lewis. "The whale optimization algorithm," *Adv Eng Softw*; Vol.95, pp.51–67, 2016.
33. M. Aziz, A.A. Ewees, A.E. Hassanien. "Whale optimization algorithm and Moth flame optimization for multilevel thresholding image segmentation," *Expert Syst Appl*; Vol. 83, pp. 242–256, 2017.

34. B.M. Ivatloo, M.M. Dalvand, A. Rabiee. "Combined heat and power economic dispatch problem solution using particle swarm optimization with time varying acceleration coefficients," *Electric Power System Res*; Vol. 95, pp. 9–18, 2013.
35. P.K. Roy. "Teaching learning based optimization for short-term hydrothermal scheduling problem considering valve point effect and prohibited discharge constraint," *Int J Electr Power Energy Syst*; Vol. 53, pp.10–9, 2013.
36. M. Basu. "Group search optimization for combined heat and power economic dispatch," *Electr Power Energy Syst*; Vol. 78, pp. 138–147. 2016.

Chapter 7

Integrated modeling and fabrication of electric vehicles for multi-criteria framework-based racing frontier ecosystems

Dharmbir Prasad, Rudra Pratap Singh, Sunetra Bhomik, Koushik Sarkar, Ranadip Roy, and Jatin Anand

INTRODUCTION

One of the biggest problems that humanity is facing is climate change. The Intergovernmental Panel on Climate Change estimates that, in the absence of significant decarbonization, the increase in global temperature would reach 1.5°C–2°C by the end of the twenty-first century. We are already experiencing its effects. One of the primary industries contributing to greenhouse gas emissions is transportation. Fifteen percent of greenhouse gas emissions globally, 27% in the US, and 42% in California are caused by transportation. In California, where this effort is centered, light-duty (passenger) automobiles alone are responsible for 28% of all greenhouse gas emissions [1]. The vehicle was designed and constructed as per standard engineering practices. The car has four non-straight-ahead wheels, an open cockpit, one seat, and open wheels (a formula-type body). From the side, there is no obstruction to the wheel/tire assembly [2,3]. No component of the car is in a keep-out zone here, and the steering is straight ahead. The following components make up the fundamental structure.

a. **Front-Head and Hoop Structure**: Moving on to the front bulkhead design, this planar structure forms the forward plane of the chassis and shields the driver's feet when viewed from the front.. Its supporting structure separates the front bulkhead from the top of the upper side impact structure to the bottom of the front hoop on the side of the chassis [4]. The front hoop is a roll bar that is close to the steering wheel and situated over the driver's legs. Similarly, the main hoop is a roll bar that is situated next to or immediately behind the driver's body.

DOI: 10.1201/9781003628279-7

b. **Chassis and Cockpit Assembly**: The chassis is the key component of any vehicle, and this engineered structural assembly holds up every working vehicle system. This assembly may consist of a single welded structure, several welded structures, or both welded and composite structures [5]. The configuration of chassis elements is projected onto a plane and only encounter compressive or tensile forces when a co-planar load is applied to any node in any direction. The optimal design of the cockpit is significant for the driver's safety. In this cockpit, the top of the racing car, the floor closeout, the interior side of the bodywork, the front bulkhead, and the firewall constitute this space, which accommodates the driver.
c. **Impact and Rollover Protection**: Thereafter, the impact attenuator is a flexible energy-absorbing structure placed in front of the front bulkhead. However, the roll hoop refers to both the front hoop and the main hoop [6]. The roll hoop bracing is a framework connecting a roll hoop to its bracing support. For unintentional protection, the primary structure's envelope and any attached secondary structures that fulfill the rollover protection envelope are needed. The section of the side of the chassis between the front hoop and the main hoop and from the chassis floor to a specific clearance height is known as the side impact structure. The volume formed by the top of the roll bar, the outside edges of the four tires, and the rollover protection envelope are united to form the surface envelope.

STUDY OBJECTIVE

In contemporary times, the strive for achieving higher and higher performances in the field of motor racing has always led to the later implication of the developed technologies initially meant for racing into passenger vehicles. The approach is quite similar but with the extra challenge of being completely electric-propelled offered us entirely new domains of technologies to explore and exploit. The following are the study goals in this work.

- To instigate a drive for design and actual manufacturing among the students at the undergraduate level in the field of EV.
- To identify the specific systems where maximum design and production time has to be focused.
- To thrive toward the most feasible design before initiating the manufacturing process and design for manufacturing.
- Achieving the most cost-effective solution with the maximum number of in-house designed and manufactured components.

LITERATURE SURVEY

Electric vehicles (EVs) have actually existed before internal combustion vehicles, with the earliest form of it being the Detroit Electric car which existed parallelly with the launch of Ford's Model-T, the vehicle that revolutionized not only the prospects of marketing the vehicle to the masses but also the assembly line being introduced for the first time [6]. But the limitations of the battery and electrical propulsion technology available at that time made practical applications of EVs uncompetitive compared to the internal combustion vehicles [7]. But with the recent advancements in the field of battery technology, specifically a boost in the energy density of the cells, and at-hand availability of Li FePo cells along with the electric propulsion system advancement have led the EV movement to acquire appreciable momentum once again [8,9]. In recent years, we have seen a major development in our country with various automotive giants thriving toward electric mobility systems of their own. Notable advancements have been showcased by the EV program of Tata Motor and Mahindra Electric Mobility Limited (formally known as Reva Motors) [10]. Also, in the public transport sector, the E-Rickshaw boom has made a significant impact on how people think about the feasibility of electric propulsion; to us, it is a revolution [11,12]. Also, with the passage of time, EVs will, not immediately, but definitely, become the economical method of transportation available. Our objectives are to be partakers and contributors in the field of EV development and manufacturing, by designing and building an EV capable of reaching high speeds, with a reasonable range and integration of in-house developed and manufactured components designed completely at the institute.

A racing EV subsystem can be divided into two categories: mechanical systems and electrical systems [13]. The design of the vehicle's subsystems has been decided upon by considering a variety of designs, solutions, and approaches. Every design style has advantages and costs of its own [14]. Even though each of these ideas complies with the mandatory specifications provided by the competition rules, they are all packed with distinctive features. The wheelbase of the vehicle, or the distance measured from the axis of the front to the back wheel, must generally be at least 60 inches. In order to establish the correct steering geometry for the car, a team also has a track width, which is the distance measured between the center of the left and right tires. The track width is usually bigger in the front than the rear. The wheelbase to front track width ratio of the vehicle varies from 1.3 to 1.7.

In the transportation sector, EVs offer many societal benefits. In 2008, 18 automobile manufacturers offered 72 models of EVs for sale in the US. Now, the Indian government has taken some initiatives regarding this matter [15]. Today's EVs run only on electricity powered by rechargeable battery units and typically drive 100 miles while consuming only 25–40 kWh of electricity. Most EVs have a range of 60–120 miles, which covers 99% of household vehicle trips according to the 2009 National Household Travel Survey. These facts prove the feasibility of EVs in social life [16,17].

When used properly, EVs produce no air pollution [18]. Whether a vehicle is moving or not, they also often produce less noise pollution than those with internal combustion engines. Electric motors do not need oxygen to burn fuel, and they also do not release any harmful gases that could contribute to global pollution [19]. The energy used by hybrid and EVs is typically produced in ways that have an influence on the environment [20–22]. However, there would be a large overall environmental gain from EV adoption.

Financial feasibility is one of the important parts of any project. The demand in the market for products or services is very much important [23,24]. As stated by the Minister of Road Transport and Highways of India, Mr. Nitin Gadkari, some initiatives have been taken to make all EVs ply on Indian roads by 2030. The West Bengal government has also made all the E-rickshaws fully tax-free to promote this [25]. That is why in this field, designing and manufacturing something very intellectual is feasible, as we already know its high demand across the world. To make something in this field and to grow the knowledge behind all these topics is financially feasible.

Designing Principle: During braking, there is a dynamic load shift from the rear tires to the front tires, due to this, the axle loads change. After calculating the axle loads, these loads can be divided by two to achieve the vertical reaction force the road conveys on the tires. Now using the frictional coefficient of the road tire interface, we can achieve the max torque exerted by the tire on the road or vice versa, and this can be equated with the torque generated by the caliper and brake disk assembly, resulting in the braking force. If the coefficient of friction between the brake pad and the brake disk is known, we can calculate the caliper force. It is simple to map the caliper force in terms of pressure to the master cylinder actuation force, which can actually be converted using the mechanical advantage of the brake pedal as the pedal effort. The general specification of the vehicle gives an overview of the vehicle in a nutshell. This includes the specification any layman can understand, and the specifications are illustrated in in Table 7.1.

MATHEMATICAL MODELING

The energy source for this study was a 2.88-kWh lithium-ion battery, which covers 48 V, 350 A, 60 Ah supply of Samsung INR 18650-25R battery. The number modules that make up the battery pack structure contains two 312 cells. The following formula is used to calculate a battery's output voltage as shown in (7.1) [26]:

Table 7.1 Proposed racing electric vehicle specification

Parameters	*Specifications*
Dry weight	145 kg
Wheelbase	1656 mm
Track width front	1648.83 mm
Track width rear	1213.704 mm
Drag coefficient	0.53
Front area	0.4 m^2
Max motor torque (each)	68.7 Nm
Max motor RPM (each)	7000
Peak/nominal motor power (each)	13.25 kW/25.2 kW
Number of cells	720 cells 18650 LiFePO
Cells configuration	30 S 3 P
Max/nominal voltage	126 V/111.11 V
Available energy	6.7 kwh
Max. braking deceleration	8.97 m/s^2
Max. longitudinal acceleration	To be tested
Time: 0–100 km/hour	3.55 s (estimated)
Top speed	158 kmph (estimated)

$$V = V_0 - RI + V_i \tag{7.1}$$

where V is the output voltage, I is the internal current, R is the internal resistance, V_0 is the open-circuit voltage, and V_i is the input potential.

In this study, a brushless direct current (BLDC) motor equipped with a hall sensor is used. Static performance scenarios in linear space are applied to the power train. When the flow time is sufficiently quick in relation to the dynamics of the system, this model can be used for dynamic simulation. The minimum consumption of the driver determines the necessary torque for the electric motor. Equation (7.2) shows how limits are related to one another [26]:

$$T_{\min} \geq T_{\lim} \geq T_{\max} \tag{7.2}$$

where $T_{\min}$ and $T_{\max}$ are negative and positive moments associated with user-defined variables depending on the operational point, respectively.

The output torque of (7.3) at port 2 of the restricted torque $T_{\lim}$ is computed using the first-order delay [26]

$$T_m = 11 + t_r \cdot s \cdot T_{\lim} \tag{7.3}$$

where the user-described constant parameter t_r is represented.

The mechanical strength is calculated using equation (7.4) [26]

$$P_{\text{mec}} = T_m \times \omega \tag{7.4}$$

where T_m is the operating torque and ω is the rotational velocity.

However, wasted electricity is calculated using equation (7.5) [26]:

$$P_{\text{lost}} = (1 - \eta) \times P_{\text{mec}} \tag{7.5}$$

The following variables affect a power train or motor's performance.

VALIDATION OF THE PROPOSED EV DESIGN

a. **Battery Accumulator Capacity Sizing**: The proposed EV model's power train requires two accumulator modules with both side pods left and right. Their technical specification has been given in Table 7.2 and Figure 7.1.
b. **Actualization of the Suspension Geometry**: After checking the compliances of the suspension design parameters with the requirements of our vehicle, we begin with the parametric modeling of the vehicle in Catia V5R20 and optimize the strength-to-weight ratio of the components involved in the front and rear suspension assembly. To optimize the strength-to-weight ratio or in other words to achieve the maximum strength for the lowest value of weight possible with the available resources, an in-hand finite elemental analysis module has been used. The component design is judged primarily upon two factors: von Misses stress and translational displacements of the node. The sub-assemblies of the rear suspension and front suspension consist of these primary components, as shown in Figures 7.2 and 7.3.

   Front Assembly: Front uprights, spindle, wheel hub, upper and lower control swing arms.

   Rear Assembly: Rear uprights, lower and upper control arms, push rod, bell crank.

Table 7.2 Accumulator specification for the EV racing model

Parameter	*Specification*
Output voltage (nominal)	48.0V
Current requirement	350 A (480 A peak)
Capacity of one battery	60 Ah
Energy stored in one accumulator	2.88 kWh
Possible configuration (as per our design)	13S 24P
No. of cells in one accumulator	312 cells
Cell makes and model	Samsung INR 18650-25R

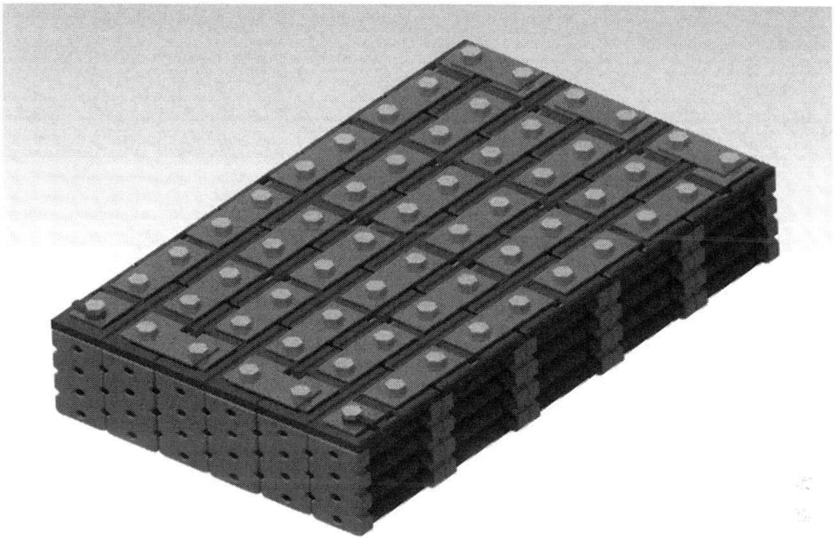

Figure 7.1 Assembled battery pack for power-train mechanism.

Figure 7.2 Assembly of the rear suspension of the vehicle.

Note: Here, all the components are not of pivotal importance; the components are prioritized on the basis of self-manufacturability and load bearing.

c. **Suspension Components Analysis:**

Front Uprights: In structural terms, these are the load transfer pathways between the spindle connecting the wheel from the hub and the mounting from the upper and lower control arms. The proposed

Figure 7.3 Assembly of front suspension with brake disk mounted onto the hub of the vehicle.

design has been illustrated below. The design has been optimized in terms of strength and also ease of manufacturability; here is where the design for manufacturing comes into play (refer to Figure 7.4).

Figure 7.4 Isometric view for front upright with the integrated spindle of the vehicle.

The uprights are designed to be manufactured into two components: the rib casing and the spindle and then finally welded into one piece. This manufacturing technique utilized saves on manufacturing time by reducing the machining time involved and the cost of the product by reducing the required raw material, which contributes to the low cost of the uprights when compared to other competitive designs which are completely milled.

The stress induced due to welding has to be eliminated to regain the strength lost due to thermal stresses induced in the uprights during integration. So, post-welding, normalization has to be used (as presented in Table 7.3). The upright has load transfer at the spindle and the supports or clamps at the attachment points via spherical ball joints to the control arm, as shown in Figures 7.5 and 7.6.

The rear uprights are also subjected to Finite Element Analysis (FEA); in this instance, they power the driven wheels and, as a result, advantages of house a bearing rather than a spindle has been shown in Table 7.4 and Figures 7.7 and 7.8.

Spindle: Since the vehicle is rear-wheel drive, spindles are the bearing surfaces that the hubs can either attach to in the case of rear uprights or use as a contact patch to allow the wheel to spin freely in the case of non-driven wheels (the front wheels in our design are non-driven wheels). The spindles are made of EN-10 carbon steel for its high surface roughness and high shear-bearing capability.

Table 7.3 Simulation output for an integrated spindle of the vehicle

Front upright	*Von Mises stress*	*Translational displacement*
Maximum value	5.85e+007 Pa	0.11 mm

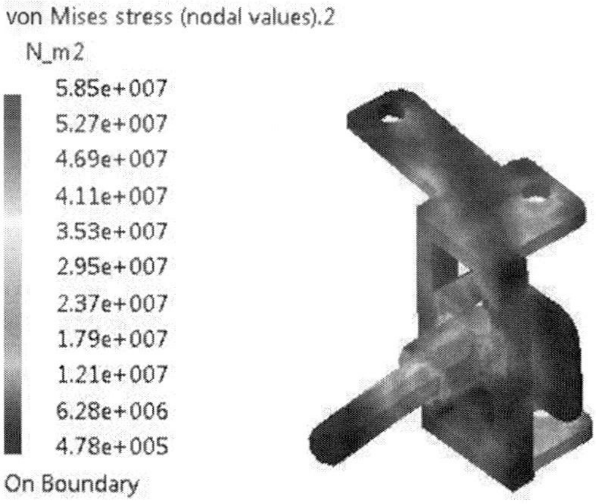

Figure 7.5 von Mises stress output for an integrated spindle of the vehicle.

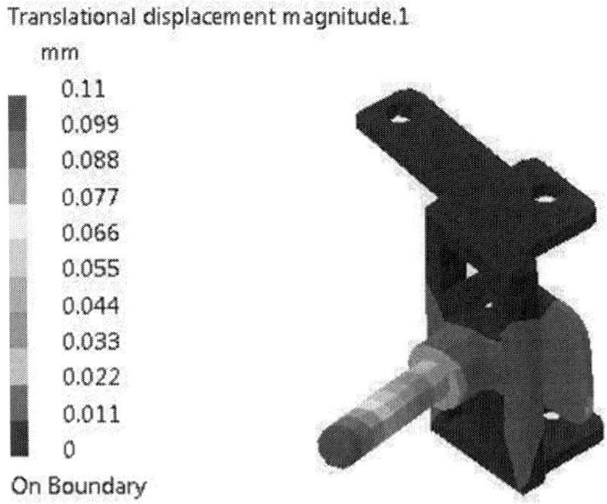

Figure 7.6 Translational displacement output for integrated spindle of the vehicle.

Table 7.4 Simulation output for rear hubs of the vehicle

Rear upright	*von Mises stress*	*Translational displacement*
Maximum value	1.81e+007 Pa	0.0112 mm

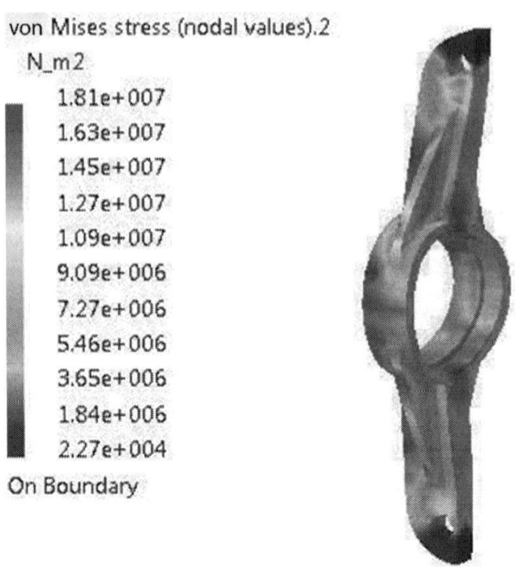

Figure 7.7 von Mises stress output for the rear upright of the vehicle.

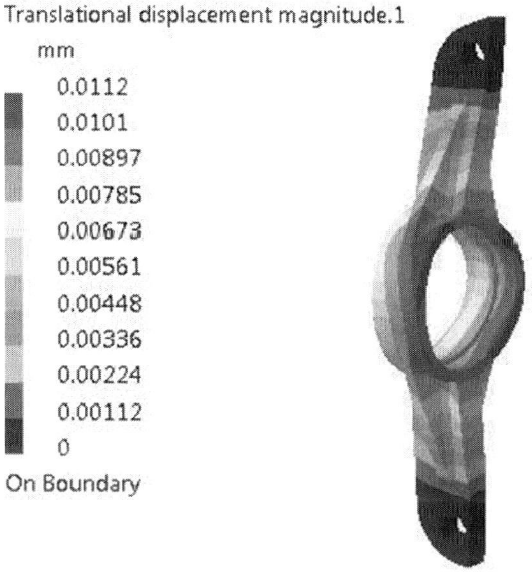

Figure 7.8 Translational displacement output for rear hubs of the vehicle.

Wheel Hubs: These components are crucial in transferring the torque and reaction forces from the wheels to the spindle. Proper methodology is used to determine the proper splines used to optimize the weight without compromising in terms of strength. The materials selected should also have a high endurance limit in the case of spindles, as they are subjected to cyclic stresses and alternating stresses, as presented in Table 7.5.

The hubs are supported at the bearing surfaces and are mainly subjected to a torsional force, and the front hubs have been analyzed with a peak torque of 68.5 Nm. The results for von Mises stress and translational displacement are shown in Figures 7.9–7.12.

Upper and Lower Control Arms: The control arms are subjected to periodically varying shock loads and bending stresses. These bending stresses constantly vary during the dynamic stages of the vehicle. Hence, the material used should not only have a high shear load-bearing capacity but also a high endurance limit (refer to Table 7.6).

A load of 700 N is applied to the end of the control arm, and its von Mises stress distribution and translational displacement are analyzed and tabulated as shown. The translational displacement is pictorially represented in the standard ISO format, as shown in Figures 7.13 and 7.14.

Table 7.5 Simulation output for the rear wheel of the vehicle

Front hub	*von Mises stress*	*Translational displacement*
Maximum value	4.22e+006 Pa	0.000535 mm

Figure 7.9 Rendering for the front hub of the vehicle.

Figure 7.10 Rendering for the rear wheel of the vehicle.

Modeling of Chassis: The chassis is the mounting and housing for all the components integrated into the vehicle. It is the framework that integrates the entire vehicle into one; chassis design has been painstakingly optimized numerous times to accommodate all the components, especially the battery accumulator and the drive motor, and to provide the driver a safe enclosure. [27] In this current design report, we present the final iteration of the space frame. The standing options from the chassis were as follows (refer to Figure 7.15).

Carbon Fiber-Reinforced Monococque: This construction methodology has significant gains in terms of superior strength-to-weight ratio and high torsional rigidity for a given cross-sectional area of material used.

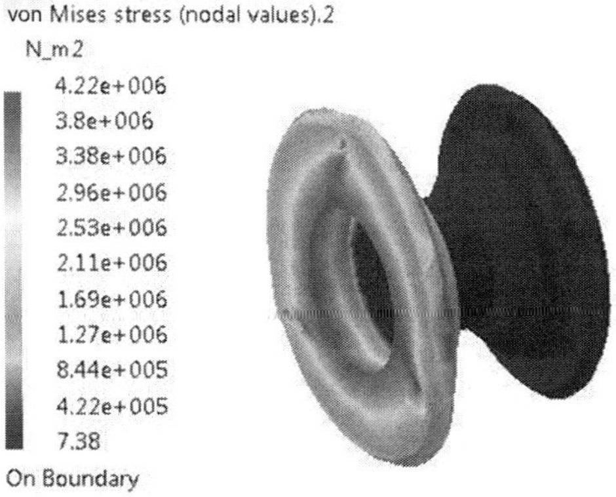

Figure 7.11 von Mises stress output for the rear wheel of the vehicle.

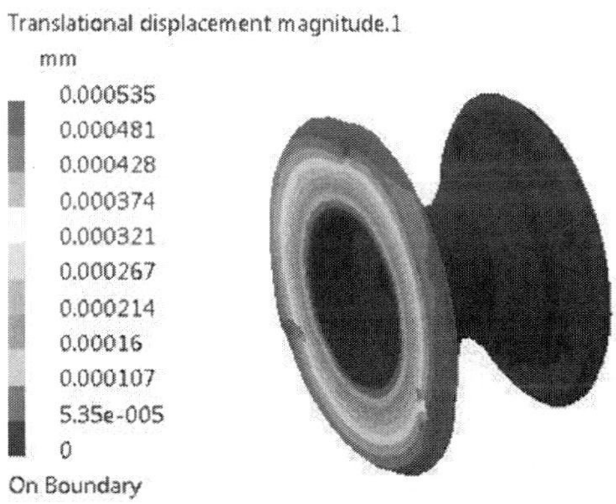

Figure 7.12 Translational displacement output for the rear wheel of the vehicle.

Table 7.6 Simulation output for the control arm of the vehicle

Control arm	*von Mises stress*	*Translational displacement*
Maximum value	7.36e+007 Pa	1.25 mm

This methodology is currently adapted in high-end motorsports vehicles along with a sandwich-like structure of CFRP (carbon fiber-reinforced plastic) and aluminum alloy (Al 7076 T) honeycomb. This methodology is not used in our case, as the manufacturing resources are highly expensive in this case.

Metal Monococque: Metal monocque, particularly sheet metal aluminum monococques, are out-ruled in our case as they have high deterioration rates in terms of endurance limits and they lose their torsional rigidity rapidly on use. The other metal monocoque is steel, which is thrice the weight of the aluminum one and will far exceed the weight limit of 30 kg for the chassis.

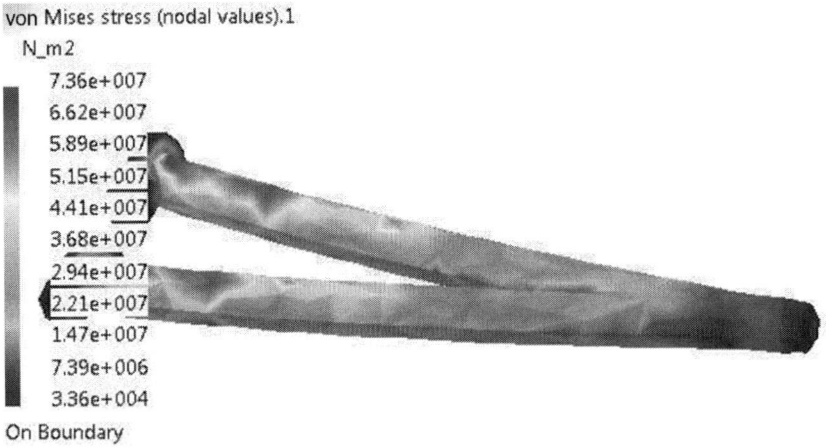

Figure 7.13 von Mises stress output for the control arm of the vehicle.

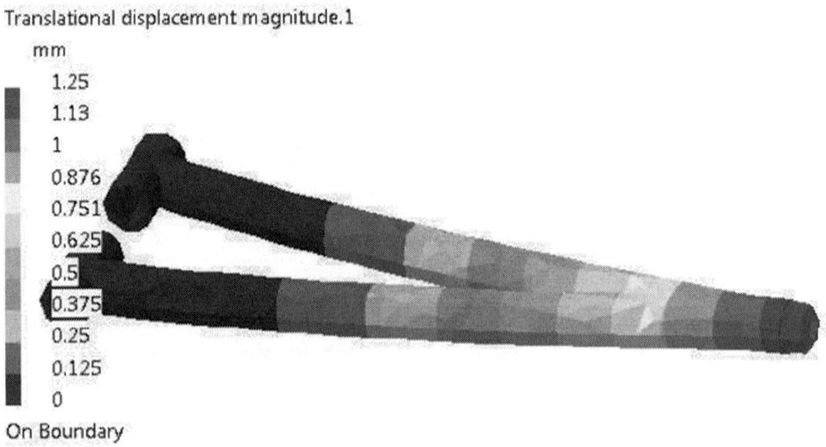

Figure 7.14 Translational displacement output for the control arm of the vehicle.

Space Frame: This method of chassis design and manufacturing is the most feasible solution in terms of strength-to-weight ratio and manufacturing resource utilization [28]. The design uses tubular metal members arranged in a three-dimensional space truss manner, such that all the loads exerted on the chassis are transferred ideally at the junction in the trusses, and the members are ideally loaded in terms of compression and tension only. The proposed frame design utilizes AISI 1541 Steel as the material, with the wall thickness of the frame members ranging from 2.0 to 2.5 mm. The chemical composition has been checked below.

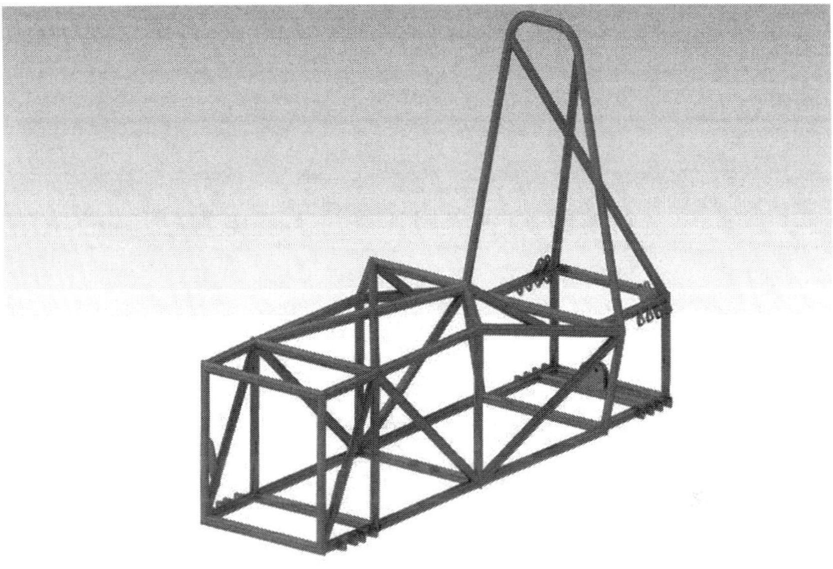

Figure 7.15 Space frame rendering of the vehicle.

The design objectives for the chassis can be concluded as follows.

1. The chassis is the component in charge of supporting all other integrated vehicle's subsystems; this is the primary unit over which all assembly will be performed. The design goals for the chassis have been envisioned as follows.
2. To create a chassis which is a complex technical tradeoff between being light and strong enough to withstand the loads imposed on it in the operating condition and static condition, with an optimal value of factor of safety (FOS).
3. All mounting locations should have a suitable, ergonomic framework that is easy to assemble. This will cut down on assembly time and eliminate the need for specialized assembly line tools, which will lower the setup costs overall.

This approach has been finalized keeping in mind the assembly and disassembly event as well as the real-world assembly line scenario.

MATERIAL SECTION

Material for the formula green chassis is medium carbon steel seamless tubes, selected for their satisfactory strength, good weldability, and general availability. The presence of chromium in the alloy makes it corrosion-resistant, with a tolerable amount of manganese in it making its machining and fabrication economy, as presented in Table 7.7. After the actualization of the frame into a parametric three-dimensional model, it was tested for three conditions:

a. Frontal Impact test for 6G collision
b. Side impact test
c. Roll over test

CASE I: FRONTAL IMPACT

The finite elemental constraint system and loads applied to the chassis are shown below, the node value for von Mises and translational displacement is graphically represented, and the maximum value and minimum value are recorded in the tabular format, as shown in Table 7.8 and Figures 7.16 and 7.17.

CASE II: SIDE IMPACT TEST

The side impact condition, causing a sudden deceleration of 5Gs and resulting in a pseudo-force being exerted upon the frame at the side impact member (SIM), is simulated in the Finite Elemental Analysis Module of Catia

Table 7.7 Material test result (as per SAE 1541)

Chemical composition		*Mechanical properties*	
C	0.411%	Tensile strength (yield)	450 MPa
Si	0.212%	Tensile strength (ultimate)	675 MPa
Mn	1.403%	% elongation	15.6%
P	0.015%	Mass density	7850 kg/m^3
S	0.005%	Hardness	NR
Cr	0.031%		
Mo, Nb, Cu, Co, B, Ti, V, W, Sn, Sb, Te, Zr	0%		
Ni	0.043%		
Al	0.029%		
Pb	0.006%		
Ca	0.002%		
Fe	97.842%		

Table 7.8 Translational displacement upon frontal impact 6G collision

6G frontal impact	*von Mises stress*	*Translational displacement*
Maximum value	8.32e+006 Pa	0.0373 mm

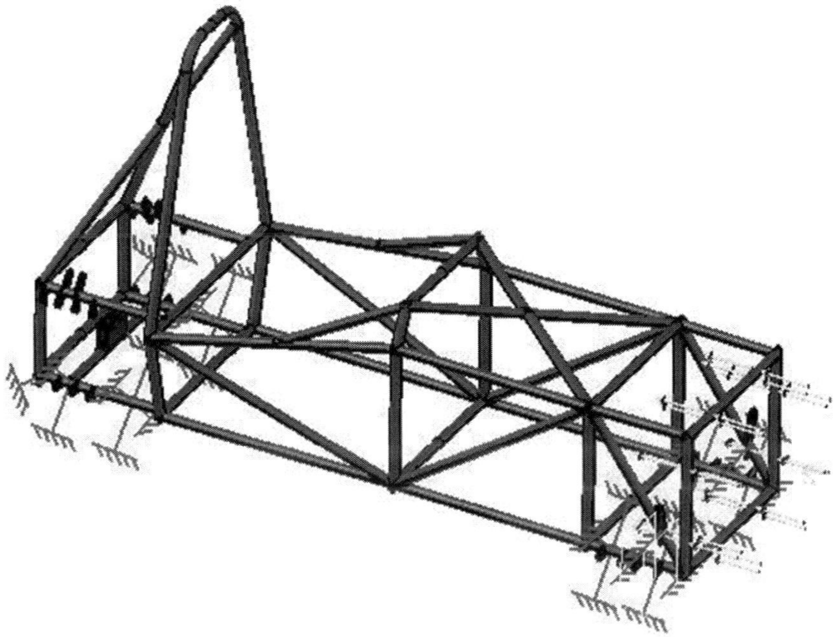

Figure 7.16 Constraints imposed on the chassis of the vehicle.

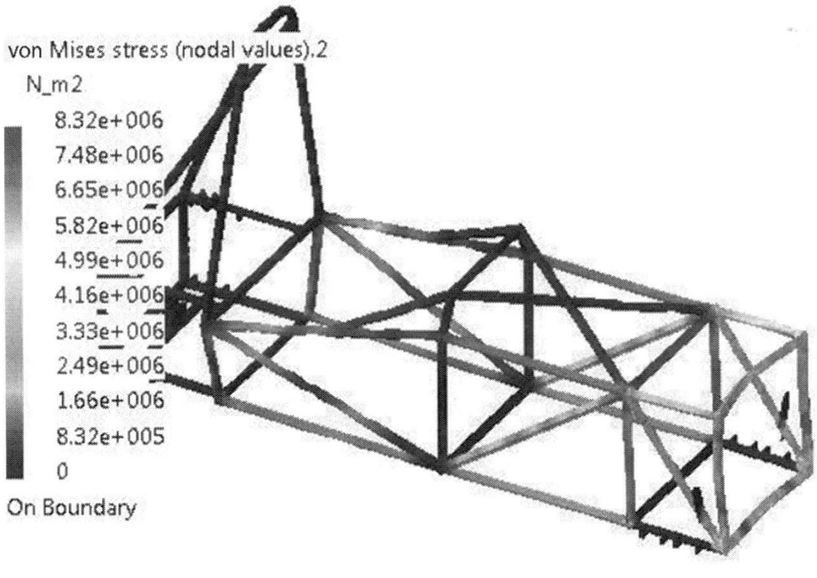

Figure 7.17 von Mises stress for frontal impact on the forward member of the vehicle.

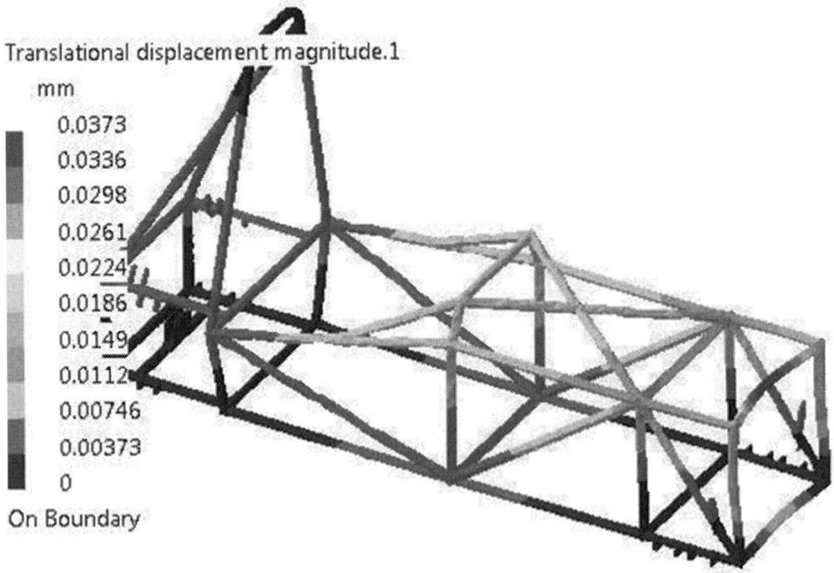

Figure 7.18 Translational displacement for frontal impact on the forward member of the vehicle.

Table 7.9 Translational displacement upon frontal impact 6G collision

5G side impact	*von Mises stress*	*Translational displacement*
Maximum value	8.59e+0007 Pa	0.541 mm

V5R20. The results are recorded and shown in Table 7.9. The force used in this case is 7350 N, as shown in Figures 7.19 and 7.20.

CASE-III: ROLL HOOP LOAD TEST OR ROLLOVER

In this case, the vehicle is assumed to roll over, and hence, the entire vehicle's mass acts over the top section of the main roll hoop member. Hence, the load applied is the weight of the vehicle along with the driver. Assuming that the driver weights 55 kg and exerting 200 kg of force on the roll hoop, corresponding result have been shown in Table 7.10 and displays in Figures 7.21–7.23.

BRAKING

The primary purpose of the braking system is to change the vehicle's kinetic energy into thermal energy, which enables the vehicle to decelerate

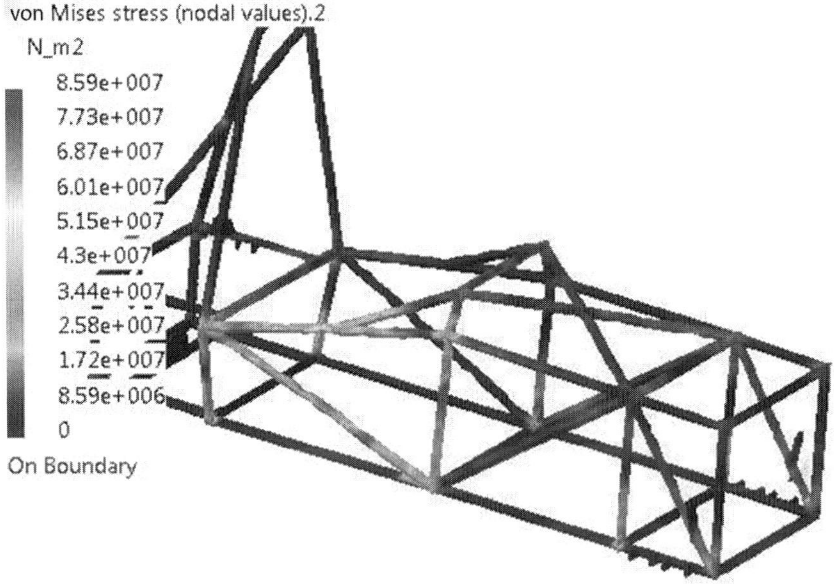

Figure 7.19 von Mises stress distribution on the side impact of the vehicle.

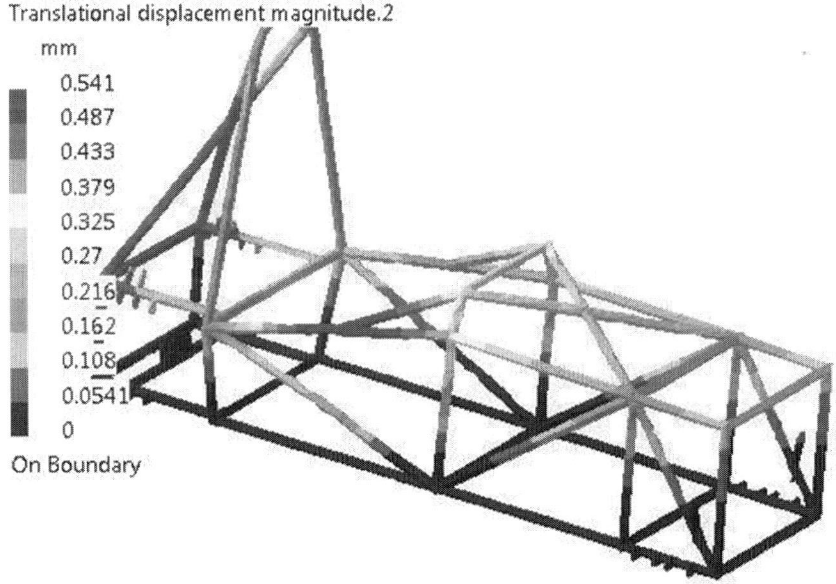

Figure 7.20 Translational displacement on the side impact of the vehicle.

Table 7.10 Simulation output for rollover test of the vehicle

Roll over test	*von Mises stress*	*Translational displacement*
Maximum value	2.43e+007 Pa	0.0795 mm

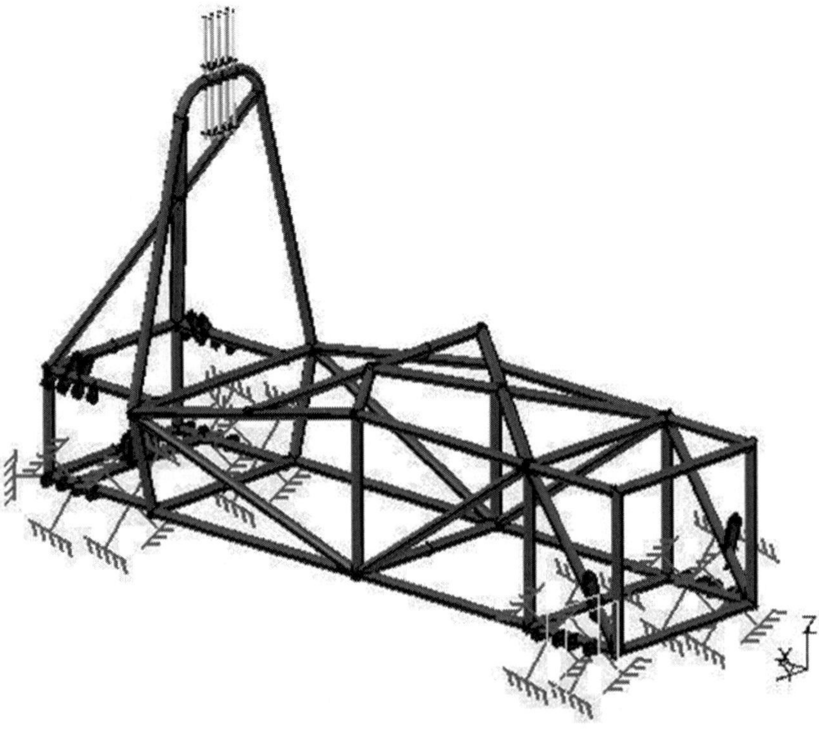

Figure 7.21 Loading and constraint conditions imposed on the frame during rollover test.

or slow down. As seen in Figure 7.24, the braking system was intended to be a hydraulic system with two master cylinders, one for each of the two front tires and one for the rear tires. An original equipment manufacturer (OEM) fluid reservoir installed locally is fastened to every master cylinder. The system's four calipers are all operated by these master cylinders.

The brake system works as follows: the driver applies force to the brake pedal, activating the master cylinder; the brake pedal then amplifies that force by providing the master cylinders with a particular amount of mechanical advantage, which causes the brake fluid to be displaced in the master cylinder. The caliper piston is then under pressure from the displaced fluid, pushing the brake pads up against the revolving braking disk. As a result, we may identify the actuation force of the caliper pistons as the output and the driver's foot force as the input. Our present design uses wood calipers with a single piston.

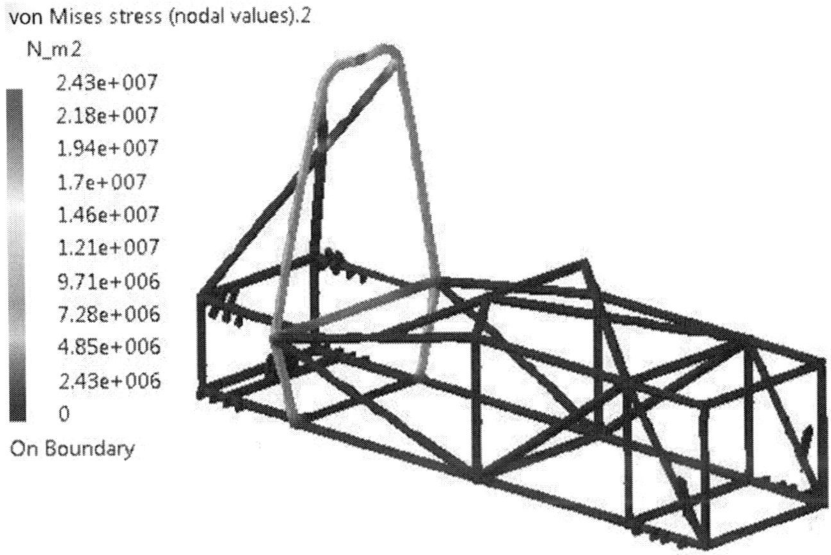

Figure 7.22 von Mises stress induced in the frame of the vehicle.

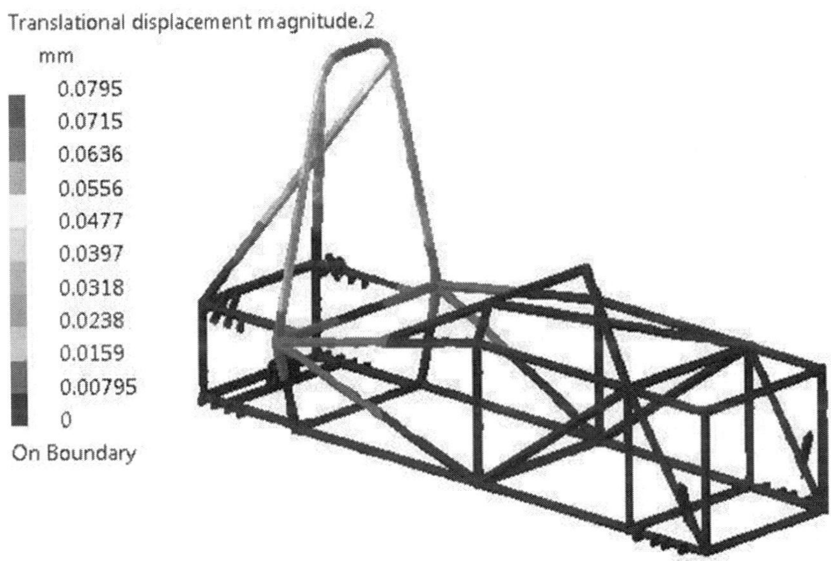

Figure 7.23 Translational displacement for rollover test of the vehicle.

The suggested braking system must be able to fully lock all four wheels of the car in the event of an emergency stop braking situation, per the competition criteria. Physically speaking, this means that the torque produced by

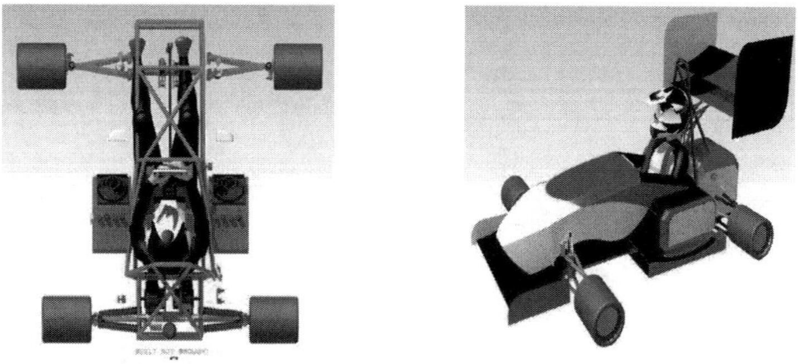

(a) top view (b) side view

Figure 7.24 Rendering of our proposed racing electric vehicle.

the caliper force applied to the brake disk must equal or exceed the force the tire applies to the road. Because of this, the braking system was designed so that the driver would need to apply a significant amount of force – without going overboard – in order to fully lock the vehicle's tires. As per the biomechanical criteria for a male in the 95th percentile of the Society of Automotive Engineers (SAE) data log, a maximum foot force of 40 N was assumed.

CONCLUSION

EV manufacturing has emerged in this decade as a result of developments in battery and power management technology, worries about rising gasoline prices, and the desire to lower greenhouse gas emissions. Despite this, the autonomy of such cars is still not sufficient, mostly because of the speed restriction. Therefore, it becomes crucial to lower the weight of the vehicle as much as possible in order to improve speed and reduce energy consumption, thereby lengthening the distance traveled between charges. Despite the crucial impact of regulations on innovation and competitiveness, little is understood about business reactions and associated consequences on performance under liberal or restrictive regulatory situations. This chapter presents the team's experience of developing and testing an EV system for a race application and plans for its future development and integration onto the vehicle.

REFERENCES

1. Hennessy, E.M. and Syal, S.M., 2023. Assessing justice in California's transition to electric vehicles. *iScience*, *26*(7), pp. 1–18.
2. Ulrich, K.T., 2005. Estimating the technology frontier for personal electric vehicles. *Transportation research part C: Emerging technologies*, *13*(5–6), pp. 448–462.

3. Liu, Z., Song, J., Kubal, J., Susarla, N., Knehr, K.W., Islam, E., Nelson, P. and Ahmed, S., 2021. Comparing total cost of ownership of battery electric vehicles and internal combustion engine vehicles. *Energy Policy*, *158*, p. 112564.
4. Lee, S.H., Yoon, B., Cho, S., Lee, S., Hong, K.M. and Suhr, J., 2023. Multidisciplinary design of door inner belt weatherstrip for simultaneous reduction of wind noise and squeaking in electric vehicles. *Materials Today Communications*, *37*, p. 107567.
5. Kwon, K., Lim, S.K., Kim, D. and Park, K., 2023. Automation program for optimum design of electric vehicle powertrain systems based on artificial neural network. *eTransportation*, *18*, p. 100267.
6. Ahmed, H.A., Megahed, T.F., Mori, S., Nada, S. and Hassan, H., 2023. Novel design of thermo-electric air conditioning system integrated with PV panel for electric vehicles: Performance evaluation. *Applied Energy*, *349*, p. 121662.
7. Aversa, P. and Guillotin, O., 2018. Firm technological responses to regulatory changes: A longitudinal study in the Le Mans Prototype racing. *Research Policy*, *47*(9), pp. 1655–1673.
8. Hou, X., Zhang, J., Ji, Y., Liu, W. and He, C., 2022. Autonomous drift controller for distributed drive electric vehicle with input coupling and uncertain disturbance. *ISA Transactions*, *120*, pp. 1–17.
9. Helveston, J.P., Wang, Y., Karplus, V.J. and Fuchs, E.R., 2019. Institutional complementarities: The origins of experimentation in China's plug-in electric vehicle industry. *Research Policy*, *48*(1), pp. 206–222.
10. Asghari, M., Al-e, S.M.J.M. and Afshari, H., 2023. Disruption management for the electric vehicle routing problem in a geographically flexible network. *Expert Systems with Applications*, *214*, p. 119172.
11. Saplinova, V., Novikov, I. and Glagolev, S., 2020. Design and specifications of racing car chassis as passive safety feature. *Transportation Research Procedia*, *50*, pp.591–607.
12. Boria, S. and Pettinari, S., 2014. Mathematical design of electric vehicle impact attenuators: metallic vs composite material. *Composite Structures*, *115*, pp. 51–59.
13. Fragassa, C., Pavlovic, A. and Minak, G., 2020. On the structural behaviour of a CFRP safety cage in a solar powered electric vehicle. *Composite Structures*, *252*, p. 112698.
14. Rong, K., Shi, Y., Shang, T., Chen, Y. and Hao, H., 2017. Organizing business ecosystems in emerging electric vehicle industry: Structure, mechanism, and integrated configuration. *Energy Policy*, *107*, pp. 234–247.
15. Kellner, Q., Hosseinzadeh, E., Chouchelamane, G., Widanage, W.D. and Marco, J., 2018. Battery cycle life test development for high-performance electric vehicle applications. *Journal of Energy Storage*, *15*, pp. 228–244.
16. Lee, Y.I., Kang, D.G., Joo, Y., Kwon, H.M., Cho, Y.S., Ryu, K.H. and Kim, M.S., 2023. Mechanical balance of plant design of lithium-air batteries for electric vehicles. *Journal of Energy Storage*, *70*, p. 107969.
17. Meng, B., Zhou, Z., Zhang, C. and Yang, F., 2023. A design method for booster motor of brake-by-wire system based on intelligent electric vehicle. *Green Energy and Intelligent Transportation*, *2*(4), p. 100110.
18. Manuguerra, L., Cappelletti, F., Manes, F. and Germani, M., 2023. A predictive eco-design method and tool for electric vehicles of Industry 4.0. *Procedia Computer Science*, *217*, pp. 248–257.

19. Diao, X., Qiu, M. and Xu, G., 2023. Electric vehicle-based express service network design with recharging management: A branch-and-price approach. *Computers & Operations Research*, *37*, p. 106469.
20. Indu, K. and Kumar, M.A., 2023. Design and performance analysis of braking system in an electric vehicle using adaptive neural networks. *Sustainable Energy, Grids and Networks*, *36*, p. 101215.
21. Er, G., Soykan, G. and Canakoglu, E., 2024. Stochastic optimal design of a rural microgrid with hybrid storage system including hydrogen and electric cars using vehicle-to-grid technology. *Journal of Energy Storage*, *75*, p. 109747.
22. Gao, X., Simeone, A. and Zhang, J., 2023. Smart decision-making for design adaptation of electric vehicles using big sales data. *Procedia CIRP*, *119*, pp. 710–715.
23. Leoncini, G., Mothier, R., Michel, B. and Clausse, M., 2023. A review on challenges concerning thermal management system design for medium duty electric vehicles. *Applied Thermal Engineering*, *236*, p. 121464.
24. Yoo, Y., Lee, H. and Lee, J., 2023. Virtual data-based generative optimization using domain-adaptive designable data augmentation (DADDA): Application to electric vehicle design. *Expert Systems with Applications*, p. 120818.
25. NewsOnAir (Aug 24, 2023). Electric vehicle market in India expected to grow to annual sales of ten million unit and create 50 million jobs by 2030, asserts, Union Minister Nitin Gadkari. https://newsonair.gov.in/News?title=Electric-Vehicle-market-in-India-expected-to-grow-to-annual-sales-of-ten-million-unit-and-create-50-million-jobs-by-2030%2C-asserts%2C-Union-Minister-Nitin-Gadkari&id=466586.
26. Hai, T., Dhahad, H.A., Jasim, K.F., Sharma, K., Zhou, J., Fouad, H. and El-Shafai, W., 2023. Deep learning-based prediction of lithium-ion batteries state of charge for electric vehicles in standard driving cycle. *Sustainable Energy Technologies and Assessments*, *60*, p.103461.
27. Khan, T., Tahir, M., Agyekum, E.B. and Yu, M., 2023. A multi-criteria design framework for sustainable electric vehicles stations. *Sustainable Cities and Society*, *101*, p. 105076.
28. Ahmed, H.A., Megahed, T.F., Mori, S., Nada, S. and Hassan, H., 2023. Performance investigation of new design thermoelectric air conditioning system for electric vehicles. *International Journal of Thermal Sciences*, *191*, p.108356.

Chapter 8

Application of artificial intelligence in wind energy generation

Chikesh Ranjan, Apurba Kumar Roy, and Kaushik Kumar

INTRODUCTION

The accelerating pace of engineering advancements has opened doors to integrating artificial intelligence (AI) across diverse energy systems. This progress has sparked innovative applications of AI within the realm of wind energy generation. One such pioneering approach revolves around strategically situating wind farms in proximity to railway tracks and harnessing AI systems to autonomously manage their operations [1]. This chapter delves into this unique synergy between AI technology and wind energy, exploring how AI's integration facilitates the strategic placement and autonomous control of wind farms near railway tracks. Wind energy stands out as a supremely clean and accessible form of energy. Its storage and utilization costs are minimal. It is imperative to reimagine this natural resource as a pinnacle of advanced technology. Relying on traditional energy extraction methods, like burning fuels and hydroelectric power, no longer aligns with our progressive ethos; we must evolve toward a more sophisticated approach, epitomizing true human advancement. Embracing cutting-edge technology is paramount in harnessing the potential of renewable energy sources, such as wind, solar, and tidal power. Despite their promise, our understanding and exploration of these resources remain rudimentary, failing to fully exploit their potential [2].

The pressing issues of pollution and electricity scarcity in rural areas are symptomatic of our global challenges. Thousands of trains and vehicles ply the roads daily, guzzling gallons of fuel with an efficiency rate of only 40%–45%. This insufficient electricity supply contributes to illiteracy and the degradation of rural communities. While solar energy adoption is prevalent in many parts of India, the focus needs to shift toward wind energy. Unlike solar panels, whose manufacturing involves hazardous chemicals and resources detrimental to the environment, wind energy emerges as a cleaner and more cost-effective alternative. The materials used in manufacturing wind turbines pose minimal environmental hazards compared to those involved in solar energy production [3,4]. With a commitment to

DOI: 10.1201/9781003628279-8

preserving the environment and promoting renewable energy, wind energy emerges as the more sustainable choice.

Trains, often propelled by strong gusts of wind, traverse the landscape at considerable speeds, serving countless villages along their routes. However, these communities face daily challenges, including limited access to electricity, education, and necessities. While reducing fuel consumption in transportation may not be immediately feasible, enhancing system efficiency offers a viable pathway forward [5].

Upon closer examination, it becomes evident that the energy derived from our current fuel sources falls short of its true value. Furthermore, the lack of awareness regarding the significance of wind energy exacerbates the issue, particularly in India. We must recognize and prioritize the potential of natural wind energy to drive sustainable progress.

Wind energy: India and the world

Wind energy has become increasingly significant both in India and globally as a renewable energy source. Here's a brief overview of its status in India and around the world [6,7].

INDIA

- **Installed Capacity:** India has made significant strides in wind energy deployment. India's installed wind energy capacity stood at over 40 gigawatts (GW), making it the fourth-largest wind power producer in the world.
- **Government Initiatives:** The Indian government has implemented various policies and initiatives to promote wind energy, including incentives, subsidies, and targets for renewable energy capacity addition. The country aims to achieve 175 GW of renewable energy capacity by 2022 and has set a target of 450 GW by 2030, with wind energy playing a crucial role in meeting these goals.
- **Challenges and Opportunities:** Despite its progress, India faces challenges such as grid integration issues, land availability, and variability in wind resources. However, technological advancements, supportive policies, and increasing investments present opportunities for further growth in the wind energy sector [8].

GLOBAL PERSPECTIVE

- **Leading Countries:** Several countries around the world have invested significantly in wind energy. China leads the global wind power market, followed by the United States, Germany, and India. These countries have large installed capacities and continue to expand their wind energy infrastructure.

- **Offshore Wind**: Offshore wind energy has gained traction in various regions, particularly in Europe and East Asia. Countries like the United Kingdom, Denmark, and Germany have pioneered offshore wind projects, tapping into strong and consistent marine winds to generate electricity.
- **Cost Competitiveness**: The cost of wind energy has declined significantly over the years, making it increasingly competitive with conventional energy sources in many parts of the world. Technological advancements, economies of scale, and favorable policies have contributed to this cost reduction.
- **Climate Goals**: Wind energy plays a crucial role in global efforts to mitigate climate change and reduce greenhouse gas emissions. Many countries have committed to transitioning to renewable energy sources as part of their climate targets under the Paris Agreement, driving further investments in wind power.

Wind turbine and its components

A wind turbine is a device that converts the kinetic energy of wind into electrical energy through the rotation of its blades [8]. Here's a brief overview of its main components [9].

- **Tower**: The tower is the structure that supports the entire wind turbine, elevating the rotor and blades to capture higher wind speeds. Towers can vary in height depending on the size and design of the turbine.
- **Rotor**: The rotor consists of multiple blades attached to a central hub. When the wind blows over the blades, it causes them to rotate. The rotor is designed to capture as much wind energy as possible and transfer it to the generator [10].
- **Blades**: Wind turbine blades are aerodynamically designed to efficiently capture wind energy. They come in various shapes and sizes, optimized for different wind conditions. Modern blades are typically made of fiberglass or composite materials for strength and durability.
- **Hub**: The hub connects the rotor blades to the main shaft. It allows the blades to rotate around their axis as they capture wind energy.
- **Generator**: The generator is located at the top of the wind turbine tower and is connected to the main shaft. As the rotor blades turn, they spin the generator, which converts mechanical energy into electrical energy through electromagnetic induction.
- **Nacelle**: The nacelle is a housing located behind the rotor hub that contains the generator, gearbox, and other essential components. It

protects these components from the elements and provides access for maintenance and servicing.

- **Gearbox:** In some wind turbines, especially older models, a gearbox is used to increase the rotational speed of the generator shaft to optimize electricity generation. However, many modern turbines utilize direct-drive systems, eliminating the need for a gearbox.
- **Controller and Sensors:** Wind turbines are equipped with sensors and control systems that monitor various parameters, such as wind speed, direction, and turbine performance. These systems optimize the turbine's operation for maximum energy output and protect it from damage in adverse conditions.

TYPES OF WIND FARMS

Wind farms are installations designed to harness the power of wind to generate electricity on a large scale [11]. There are several types of wind farms, each with its characteristics and suitability for different environments.

- **Onshore Wind Farms:** These are the most common type of wind farms located on land. Onshore wind farms typically consist of rows of wind turbines installed in open areas with consistent wind flow. They are relatively easier to construct and maintain compared to offshore wind farms, making them more cost-effective in many cases.
- **Offshore Wind Farms:** Offshore wind farms are situated in bodies of water, usually in coastal areas or offshore locations. They utilize larger turbines mounted on platforms anchored to the seabed. Offshore wind farms have the potential to generate higher amounts of electricity due to stronger and more consistent wind speeds, but they also entail higher installation and maintenance costs.
- **Nearshore Wind Farms:** Nearshore wind farms are positioned close to the coastline, often within a few kilometers of the shore. They combine some of the advantages of onshore and offshore wind farms, such as relatively easier installation and maintenance compared to offshore installations, while still benefiting from strong and consistent coastal winds.
- **Floating Wind Farms:** Floating wind farms are a relatively newer technology that involves placing wind turbines on floating platforms anchored to the seabed in deeper waters. This allows for the deployment of wind turbines in locations where traditional fixed-bottom offshore installations are not feasible. Floating wind farms have the potential to access even stronger winds further offshore.
- **Hybrid Wind Farms:** Hybrid wind farms integrate wind energy with other forms of renewable energy generation, such as solar or energy storage systems. These integrated systems help to stabilize energy output and maximize overall efficiency, particularly in areas with variable weather conditions.

Each type of wind farm has its own advantages and challenges, and the choice of technology depends on factors such as wind resources, geographical location, environmental considerations, and economic feasibility.

IMPACT OF WIND POWER NEAR RAILWAY TRACKS

The impact of wind power projects near railway tracks can be multifaceted, influencing various aspects such as energy generation, transportation, and local communities. Here are some potential impacts.

- **Energy Generation**: Placing wind power projects near railway tracks can leverage the consistent wind flow generated by passing trains. This proximity allows wind turbines to capture the kinetic energy from the moving trains more efficiently, potentially increasing energy output [12–15]. As a result, wind power projects near railway tracks may yield higher electricity generation compared to installations in other locations with less consistent wind patterns.
- **Synergy between Infrastructures**: Integrating wind power projects with existing railway infrastructure demonstrates a synergistic approach to resource utilization. By utilizing available land adjacent to railway tracks, wind turbines can be strategically positioned to take advantage of the wind energy generated by passing trains. This symbiotic relationship between wind power and rail transportation highlights innovative ways to optimize land use and infrastructure efficiency.
- **Transportation Safety and Accessibility**: Wind power projects near railway tracks must consider the safety and accessibility implications for rail operations. Proper planning and coordination are essential to ensure that wind turbines do not interfere with railway signaling, clearance requirements, or maintenance activities. Additionally, access roads and infrastructure for wind turbine installation and maintenance should be designed to minimize disruptions to railway operations and ensure the safety of railway personnel and passengers [16].
- **Community Impacts**: The proximity of wind power projects to railway tracks may affect nearby communities in various ways. On one hand, increased renewable energy generation can contribute to local economic development and job creation, potentially benefiting communities along railway routes. However, concerns such as noise pollution, visual impacts, and land use conflicts may arise and require careful consideration and community engagement during project planning and development.
- **Environmental Considerations**: Wind power projects near railway tracks should also address environmental considerations, including wildlife habitat, ecological corridors, and potential impacts on sensitive ecosystems. Proper siting and mitigation measures can help

minimize adverse effects on biodiversity and ecosystem services, ensuring sustainable coexistence between wind energy infrastructure and natural environments [17–20].

The impact of wind power projects near railway tracks depends on various factors such as project design, location, and stakeholder engagement. By considering these factors comprehensively and implementing appropriate measures, wind power projects can effectively harness wind energy while minimizing negative impacts on railway operations, communities, and the environment.

WORKING OF WIND TURBINES THROUGH AI SYSTEMS

The working of wind turbines controlled through AI systems near railway tracks involves a complex interplay of data gathering, analysis, and decision-making to optimize energy generation while ensuring safe railway operations. Here's a simplified explanation of how this process could function [21].

- **Data Gathering**: AI-controlled wind turbines near railway tracks continuously gather real-time data from various sources. This includes wind speed and direction sensors mounted on the turbines, weather forecasts, railway scheduling systems, and train detection sensors along the tracks. Additionally, turbines may also collect data on their own operational parameters such as rotor speed, power output, and health status.
- **Data Processing and Analysis**: The collected data is processed and analyzed by AI algorithms in real-time. These algorithms use machine learning techniques to identify patterns, predict future events, and make decisions based on the available information. For example, AI systems can analyze wind speed and direction data to anticipate changes in wind patterns and adjust turbine operation accordingly.
- **Train Detection and Prediction**: AI systems are trained to detect approaching trains using sensors installed along the railway tracks. By analyzing the data from these sensors and historical train schedules, the AI algorithms can predict the timing and trajectory of trains passing near the wind turbines. This information is crucial for adjusting turbine operations to maintain safe distances and minimize disruptions to railway operations.
- **Turbine Control**: Based on the analyzed data and predictions, AI systems adjust the operation of the wind turbines near railway tracks. This may involve dynamically optimizing turbine orientation, blade pitch angles, and rotational speeds to maximize energy capture while

ensuring safety and efficiency. For example, turbines may be repositioned to align with prevailing wind directions or reduce interference with passing trains.

- **Safety Protocols**: AI-controlled wind turbines implement safety protocols to mitigate risks associated with their proximity to railway tracks. This includes automatically shutting down turbines or adjusting their settings when trains are detected nearby, as well as implementing emergency shutdown procedures in case of unforeseen events or equipment failures.
- **Monitoring and Maintenance**: AI systems continuously monitor the performance of wind turbines and railway operations in real-time. Any anomalies or issues detected are promptly addressed through remote monitoring and management. Additionally, AI algorithms can schedule preventive maintenance activities based on turbine health data and operational patterns to minimize downtime and maximize reliability.
- **Adaptive Learning**: Over time, AI systems learn from past experiences and feedback to improve their performance and decision-making capabilities. This adaptive learning process allows the AI algorithms to continuously refine turbine control strategies, optimize energy generation, and enhance safety near railway tracks.

By integrating AI control systems with wind turbines near railway tracks, it is possible to optimize energy generation while ensuring safe and efficient coexistence with railway operations. This approach maximizes the utilization of renewable energy resources while minimizing environmental impacts and enhancing overall system reliability [22–24].

While controlling wind turbines through AI systems near railway tracks offers numerous benefits, several challenges and potential problems may arise. Here are some issues to consider.

- **Interference with Railway Operations**: Wind turbines near railway tracks may interfere with railway operations if not properly controlled. Turbines rotating too close to the tracks could pose safety risks, generate vibrations, or cause electromagnetic interference with signaling systems, potentially leading to accidents or disruptions in train schedules.
- **Safety Concerns**: Ensuring the safety of railway personnel, passengers, and maintenance workers is paramount. AI-controlled wind turbines must have robust safety protocols in place to detect approaching trains accurately and initiate emergency shutdown procedures when necessary. Failure to do so could result in accidents or injuries.
- **Noise and Vibration**: Wind turbines produce noise and vibrations during operation, which could impact nearby railway infrastructure and communities. AI systems should mitigate these effects by optimizing

turbine operation and implementing sound-dampening measures. However, noise and vibration levels may still exceed acceptable limits, leading to complaints or regulatory issues.

- **Visual Impact**: Wind turbines near railway tracks may alter the visual landscape and impact the aesthetic appeal of the surrounding area. This could lead to objections from local communities, particularly if the turbines obstruct scenic views or cultural heritage sites. Effective communication and stakeholder engagement are essential to address concerns and gain community acceptance.
- **Environmental Impact**: The installation of wind turbines near railway tracks may have environmental implications, including habitat disruption, bird collisions, and landscape degradation. AI systems should consider ecological factors when controlling turbine operations to minimize adverse effects on wildlife and ecosystems. Environmental impact assessments and mitigation measures may be necessary to address these concerns.
- **Maintenance Challenges**: Accessing and maintaining wind turbines near railway tracks can be challenging due to limited space and logistical constraints. AI-controlled predictive maintenance strategies can help anticipate equipment failures and optimize maintenance schedules. Still, servicing turbines near railway operations requires careful planning and coordination to ensure safety and minimize disruptions.
- **Regulatory Compliance**: Wind turbines near railway tracks must comply with regulatory requirements and safety standards established by relevant authorities. AI-controlled systems should adhere to applicable regulations governing wind energy installations, railway infrastructure, and environmental protection. Failure to comply with regulatory requirements could result in fines, legal liabilities, or project delays.

Addressing these challenges requires comprehensive planning, collaboration between stakeholders, and robust risk management strategies. By proactively addressing potential problems, wind turbines controlled through AI systems near railway tracks can maximize energy generation while minimizing risks and ensuring the safety and well-being of all stakeholders involved.

EXPERIMENTATION

We proposed a novel and innovative concept for capturing energy using a new type of windmill. This windmill can be adjusted to maintain a consistent output while achieving the desired energy production.

The power generated by a wind turbine is given by

$$P = \frac{1}{2} \rho a v^3$$

where,

P = power in watts

a = area (front) in m^2

$\frac{1}{2}$ = Betz coefficient

V = velocity of gust in m/s

ρ = density of air = 1.26 kg/m^3

We cannot change the turbine's radius, so we will adjust the height of the vertical turbine. Keeping the radius constant at 0.3 m (30 cm), and considering an average train speed of 15.27 m/s (55 km/h), air density of 1.26 kg/m^3, and a Betz coefficient of 0.5, the turbine height should range between 30 and 55 cm.

This height range ensures that the turbine remains compact and less bulky, as shown in Figure 8.1, facilitating easy installation along railway tracks. Increasing the height allows for greater air intake, thereby enhancing output compared to altering the radius.

In a wind tunnel setup for testing the vertical turbine, the turbine is placed in the test section where uniform airflow is ensured using flow straighteners. The wind source, typically fans or blowers, generates adjustable airspeeds to replicate the average train speed of 15.27 m/s. The setup includes instrumentation like anemometers, pressure sensors, and force transducers connected to a data acquisition system to record measurements. The testing procedure involves calibrating sensors, taking baseline measurements, and conducting incremental tests by varying wind speeds and turbine heights

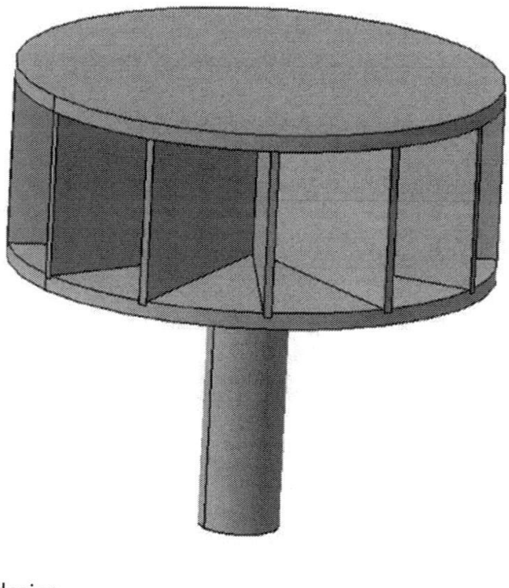

Figure 8.1 Turbine design.

between 30 and 55 cm. Data analysis focuses on performance curves, identifying the optimal turbine height, and evaluating efficiency. This controlled testing helps determine the most effective turbine configuration for installation along railway tracks (Figure 8.2).

In this setup, the turbine is placed centrally in the test section. The fans or blowers are located at the entrance to generate airflow, and flow straighteners ensure the air moves uniformly. Sensors positioned strategically within the test section measure various parameters, while the data acquisition system records and analyzes these measurements to evaluate the turbine's performance under different conditions.

Figure 8.3 illustrates that by maintaining a constant turbine diameter of 60 cm and a height of 45 cm, the power generated varies with different wind velocities, ranging from 150 to 620 W.

Figure 8.2 Wind tunnel setup for testing of the turbine.

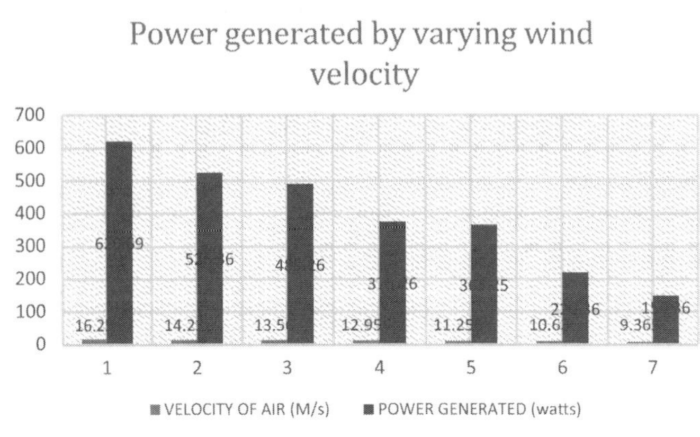

Figure 8.3 Plot of power generated by varying wind velocity.

CONCLUSION

The world is advancing rapidly, yet we still rely heavily on non-renewable energy sources. Wind energy, a renewable resource, can be harnessed using wind turbines, and we propose a new turbine design to capture gusts of wind generated by passing trains. AI is revolutionizing the wind energy sector by enhancing efficiency, reliability, and environmental sustainability. As AI technology progresses, we foresee even more ground-breaking advancements, solidifying wind energy's role as a cornerstone of sustainable power generation. By integrating AI into tasks like turbine control, predictive maintenance, and grid optimization, wind farms can maximize energy output, minimize downtime, and adapt to changing conditions in real-time. However, to fully realize AI's potential in wind energy, challenges like data quality, algorithm robustness, and ethical considerations must be addressed. Through continuous research, development, and collaboration, AI can drive the transition toward a cleaner, more resilient energy future powered by wind.

FUTURE SCOPE

The future scope of integrating AI into wind energy generation is poised for transformative growth, promising to revolutionize efficiency, reliability, and sustainability in renewable energy production. Advancements in AI-driven control systems, predictive analytics, and machine learning techniques will enable wind farms to achieve unprecedented levels of energy optimization, operational autonomy, and grid integration. Furthermore, the convergence of AI with other emerging technologies, such as the Internet of Things (IoT) and renewable energy storage, will unlock synergies and drive interdisciplinary innovation. As AI continues to evolve, cross-sector collaboration and investment will be essential to realize the full potential of AI-driven solutions in shaping the future of wind energy generation, paving the way for a cleaner, more resilient, and interconnected energy landscape.

REFERENCES

1. A. Dilimulati, T. Stathopoulos, M. Paraschiovoiu, Wind turbine designs for urban applications: a case study of shrouded diffuser casing for turbines, *J. Wind Eng. Ind. Aerod.* 175 (2018) 179–192.
2. A. Englberger, A. Dornbrack, Impact of the diurnal cycle of the atmospheric boundary layer on wind-turbine wakes: a numerical modelling study, *Inst. Phys. Atmos.* 166 (2016) 423–438.
3. A. Kareem, Emerging frontiers in wind engineering: computing, stochastics, machine learning and beyond, *J. Wind Eng. Ind. Aerod.* 206 (2020). 104320.

4. A. Stetco, F. Dinomohammadi, X. Zhao, V. Robu, D. Flynn, M. Barnes, J. Keane, G. Nenadic, Machine learning methods or wind turbine condition monitoring: a review, *Renew. Energy* 133 (2019) 640–635.
5. A. Winstral, T. Jonas, N. Helbig, Statistical downscaling of gridded wind speed data using local topography, *Am. Meteorol. Soc.* 18 (2017) 335–348.
6. P. Evans, The psychology of climate change skeptics, and building a better wind turbine, *CBC News*, 30th November 2018.
7. H. Zhou, Y. Lue, X. Liu, R. Chang, B. Wang, Harvesting wind energy in low-rise residential buildings: design and optimization of building forms, *J. Clean. Prod.* 167 (2017) 306–316.
8. H. Demolli, A.S. Dokuz, A. Ecemis, M. Gokcek, Wind power forecasting based on daily wind speed data using machine learning algorithm, *Energy Convers. Manag.* 198 (2019) 111823.
9. J. Maldonado-Correa, J.C. Solano, M. Rojas-Moncayo, Wind power forecasting: a systematic literature review, *Wind Eng.* 45 (2019) 1–14.
10. K. Klink, Trends in mean monthly maximum and minimum surface wind speeds in the coterminous United States, 1961 to 1990, *Clim. Res.* 13 (1999) 193–205.
11. K. McInnes, T. Erwin, J. Bathols, Global climate model projected changes in 10 m wind speed and direction due to anthropogenic climate change, *Atmos. Sci. Lett.* 12 (2011) 325–333.
12. M. Begam, S.N. Deepa, Optimized nonlinear neural network architectural models for multistep wind speed forecasting, *Comput. Electr. Eng.* 78 (2019) 32–49.
13. M. Fischetti, M. Fraccoaro, Machine learning meets mathematical optimization to predict optimal production of offshore wind parks, *Comput. Oper. Res.* 106 (2019) 289–297.
14. P. Newman, Cool planning: how urban planning can mainstream responses to climate change, *Cities* 103 (2020) 102651.
15. P. Saeed, The importance of renewable energies with emphasize on wind power, *Int. J. Eng. Res. Technol.* 9(6) (2020) 164650–164666.
16. S. Harbola, V. Coors, One dimensional convolutional neural network architectures for wind prediction, *Energy Convers. Manag.* 195 (2019) 70–75.
17. S. Pryor, R. Barthlemie, Climate change impacts on wind energy: a review, *Renew. Sustain. Energy Rev.* 14 (2009) 430–437.
18. S. Shamshirband, T. Rabczuk, K.-W. Chau, A survey of deep learning techniques: applications in wind and solar energy resources, *IEEE* 7 (2019).
19. T. Blanchard, B. Samanta, Wind speed forecasting using neural network, *Wind Eng.* (2019) 1–16.
20. T. Stathopoulos, H. Alraweshdeh, A. Al-Quraan, B. Blocken, A. Dilimulati, M. Paraschivoiu, P. Pilay, Urban wind energy: some views on potential and challenges, *J. Wind Eng. Ind. Aerod.* 179 (2018) 146–157.
21. Y. Ozmen, E. Baydar, J. Van Beeck, Wind flow over the low-rise building models with gabled roofs having different pitch angles, *Build. Environ.* 95 (2016) 63–74.
22. Y. Tominaga, S-i Akabayashi, T. Kitahara, Y. Arinami, Air flow around isolated gable-roof buildings with different roof pitches: wind tunnel experiments and CFD simulations, *Build. Environ.* 84 (2015) 204–213.

23. Z. Tasneem, A. Al Noman, S. Das, D. Saha, R. Islam, F. Ali, F." Badal, H. Ahamed, S. Moyeen, F. Alam, An analytical review on the evaluation on wind resources and wind turbine application: prospect and challenges, *Dev. Built Environ.* 4 (2020) 10–33.
24. Z. Ti, X.W. Deng, H. Yang, Wake modeling of wind turbines using machine learning, *Appl. Energy* 257 (2020) 114025.

Chapter 9

Governing and empowering independent power producers in South Africa

Pfano Mashau and Jabulani Christopher Nyawo

INTRODUCTION

Africa faces a power crisis. Generating power requires significant investment and support infrastructure, which many African countries need more. Furthermore, regulations in African countries often restrict power producers' innovation and incentives to generate electricity. These restrictions include safety measures and limits on the amount of power a producer can generate. As a result, Africans needing additional and supplemental energy for their homes and businesses often resort to small solar panels and generators. While these solutions provide immediate relief during power shortages, they must be more sustainable.

As the energy problem worsens, governments increasingly turn to independent power producers (IPPs) to help meet their countries' demands. For instance, in 2022, the South African government amended its regulations to allow IPPs to supply power to the grid following continuous load shedding. This amendment creates business opportunities that must be adequately regulated to ensure broader involvement and benefits. Given the high levels of inequality in countries like South Africa, it is essential to study these regulations to manage power supply effectively and explore the opportunities this amendment offers to ordinary business-minded citizens.

This chapter reviews how countries outside Africa manage independent power production and the opportunities that arise. The literature provides a theoretical foundation for recommendations on how Africa, specifically South Africa, can effectively manage independent power generation.

DESIGN FOR THE CRITICAL

This chapter is based on a collection of peer-reviewed studies addressing energy and electricity globally. The discussion is organized as follows: the initial subsections provide context on the global energy crisis from a world perspective, followed by African and South African perspectives.

 DOI: 10.1201/9781003628279-9

The subsequent sections discuss solutions and recommendations for African countries, including South Africa.

This systematic review derives its results from data described in scholarly articles. It reviews published articles that address a specific situation using a systematic method of identifying relevant studies to minimize biases and errors [1]. The study extensively utilizes literature published after the Paris Agreement, which binds signatory countries to seek environmentally friendly energy sources to protect the planet.

The following steps guide this systematic review [2]: starting with framing questions for a review, sourcing relevant work, selecting and assessing the quality of studies, interrogating and summarizing the evidence, and lastly, interpreting the findings. This chapter utilized document analysis to collect and critically analyze the secondary data. Broadly, peer-reviewed articles within the context of Africa were consulted to understand better the issue of governing and empowering IPPs in South Africa. Furthermore, government policy documents and reports were also revisited. Although information about energy production and supply is available, more studies need to be conducted on IPPs and their contribution to the energy sector in Africa and South Africa in particular. Studies between the period 2014 and 2022 on the matter under investigation were utilized.

GLOBAL ENERGY CRISIS

Global warming has challenged the world to find sustainable energy sources and solutions to the power crisis. Nations have been switching to green energy instead of producing energy from finite resources. This transition is not simple as stated by Berahab et al. [3]; the world is dependent on fossil fuels, and there are risks of a poorly managed energy transition. In 2022, bad weather, the war in Russia, and resource pricing reminded the world to search for alternative energy sources as a matter of urgency. Alternative power sources are sustainable and cheaper than fossil fuels, gas, and coal-generated electricity. However, the transition is not straightforward because many countries have invested heavily in traditional power generation sources and are unwilling to abandon their infrastructures.

Berahab [3] states that the developed countries have achieved some energy security and are now focusing on energy transition. Berahab shows that in 2021, the demand for coal for power generation increased, leading to soaring natural gas and coal prices, which is against many countries' energy transition efforts and climate action. This indicates that countries will not abandon coal, fuel, and gas stations sooner. Instead, they remain essential sources of energy.

Analytics [4] states that governments must completely rearrange their energy supplies away from fossil fuels. In agreement with the discussion above, there is an increase in the demand for fossil gas production, pipelines, and liquefied natural gas (LNG) facilities. This indicates that the Paris Agreement to limit carbon emissions is becoming more challenging to accomplish. Research studying European industries' response to the energy crisis found that alternative energy sources are limited, and businesses tend to return to the government for subsidies to cover the high cost of energy [5].

Like many other parts of the world, China is experiencing an energy crisis and exploring alternative energy sources. Chinese researchers have started investigating how rural areas can contribute to alleviating the situation. Zheng et al. [6] studied biomass energy utilization in rural areas to mitigate the energy crisis and global warming. The study concludes that biogas production and utilization will effectively alleviate the energy crisis, including CO_2 emission, which aligns with the Chinese government's goal to promote carbon emissions. Rural areas are essential to a country's economy, and initiatives in rural areas must be sustainable and profitable. Therefore, governments need to involve rural areas in addressing national crises. In China, a study identifies promising strategies that entail livestock and biogas projects as necessary for consuming redundant agricultural residues, offering livelihood energy, the standard of living, and increasing the villagers' incomes.

China is worth observing as the world's largest primary energy consumer; its economy is large enough for its energy development strategy to influence the global energy structure and environmental conditions. Hua et al. [7] further support the role of government in supporting pro-environment behavior. Relevant government departments and communities are called to support social capital and actively guide residents to participate in environmentally friendly initiatives. Although China has abundant fossil fuel, the global environment deterioration necessitates the urgency of new pro-environment energy-generating initiatives. Pan and Dong [8] suggest the following policy considerations for China:

- The development of new energy technology and an oppositional capital orientation for new energy development,
- Technology spillover and joint regional development,
- Diversifying energy forms and improving China's overall clean energy development.

These are important considerations as they allow a country to put in infrastructure that enables energy generation, thereby creating an environment where know-how spreads among locals and provides various opportunities for participation.

A study providing a comparative analysis of several countries also shows the influence of renewable energy and provides evidence of environmental duplication. Russia has been identified as one of the leading countries transitioning to alternative energy sources, more effectively, for further energy supply [9]. Russia notes the importance of ensuring financial sustainability and tariff equalization to maximize welfare. Electricity is the foundation of most economies in the world, and transition to unreliable and new energy sources may reduce the provision of a good standard of living. Bogodukhova et al. [9] identified the need for more active implementation of renewable energy sources and a profitable direction for developing an alternative energy system for Russia. This indicates that advanced countries are reluctant to implement alternative power source strategies to avoid the possible adverse impact that may result from the decision. The literature shows that electricity or energy is vital for economic performance.

Islam et al. [10] studied the energy situation in Bangladesh, a small island with 170 million people. It mentions that "the process of ensuring affordable, uninterruptable, and quality energy for all with limited resources and infrastructure is a great challenge to the government." The above observation is important to note as it indicates that countries of different sizes and developmental states need help securing reliable electricity and sufficient green energy. Bangladesh made the following corrective steps:

- "The adaptation of fuel diversification in the energy and power sectors' roadmap and investments unveiled the need for the right policies for energy exploration and production,
- equal energy distribution in urban and rural areas,
- removing political and bureaucratic bottlenecks,
- controlling and eliminating corruption,
- timely completion of power projects, and
- putting the right people in the right place."

These steps needed to be accompanied by actions and projects that are sustainable and renewable. The country implemented solutions such as (1) strengthening renewable energy sources (such as solar and wind as part of their green energy strategy), (2) developing more nuclear power plants to cater to the expansion of the grid, which requires the country to invest in energy-efficient technologies, and (3) creating an energy culture through social change, which includes more entrepreneurial initiatives that build the economy.

In this section, it is essential to note that the world faces a challenge that requires collective action and sustainable solutions. A study states that "the global energy crisis is caused by world population growth, an escalating increase in demand, and continued dependence on fossil-based fuels for generations" (Coyle and Simmons, 2014) [11]. This trend is global, and countries are faced with similar challenges. A study suggests that it is widely accepted that increases in greenhouse gas concentration levels, if not reversed, will result in significant changes to world climate with consequential effects on our society and economy. These changes are noticeable at the present moment, and countries are compelled to take action immediately. The following changes will illustrate the importance of considering the state of the economy when implementing the suggested solutions, particularly for Africa.

AFRICA ENERGY CRISIS

Africa faces an energy crisis, particularly in the provision of electricity, similar to the rest of the world. However, there needs to be more proper documentation and research detailing the extent of this problem from the perspective of each African country. South Africa's electricity crisis is well-documented compared to the rest of the continent. The South African issue stems from the country's aging infrastructure, which needs to supply consistent electricity. South Africa is investigating the best alternative power sources to complement coal-produced electricity. Given its abundant coal reserves, the country is also a large coal exporter to other countries. The South African government is looking into incorporating IPPs into the national grid amid severe load shedding that needs immediate attention to save the economy.

Africa's large economies face electricity crises that are as severe or worse than South Africa's. A study in Nigeria highlights that economic prosperity relies on reliable electricity, making the power or energy sector a pillar of the economy [12]. The study states that reliable access to electrical power in homes and businesses remains a significant obstacle in Nigeria. Olatunji et al. [12] conclude the following regarding Africa's solution to the energy crisis:

- Nigeria, like other countries, aspires to implement environmentally friendly energy sources. However, the progress of migrating to renewable energy in Nigeria is slow.
- The transition to alternative energy sources requires substantial investment and supporting infrastructure. Most of Nigeria's power-generating facilities operate at low efficiency. Complementing infrastructure is needed for a transmission and distribution network that allows renewable energy to contribute to the national grid.
- Countries like South Africa have considered nuclear power plants as a solution, but political and safety concerns have deterred their implementation. Similarly, Nigeria's volatile security situation makes nuclear power plants an unsuitable solution.

A study in Sub-Saharan Africa (SSA) emphasizes the importance of proper planning mechanisms [13]. The study documents that over half of the 1.2

billion people in SSA in 2020 lacked access to modern energy services and experienced an energy poverty crisis. Africa is well-positioned to harvest renewable energy resources due to its geographic advantages. However, following the Paris Agreement's goal of capping atmospheric temperature to mitigate climate change poses a significant challenge for Africa. South African economies have industrialization policies aimed at development, and it is counterproductive to expect these countries to power themselves entirely with environmentally friendly electricity sources while simultaneously achieving their developmental plans. The migration to renewable energy and economic growth will require innovative strategies and flexible policies. Musonye et al. [13] suggest that SSA must deliberately "employ advanced energy modelling expertise from developed countries to develop local energy modelling capacity, incorporating both technology-rich bottom-up and economically oriented top-down methods, while a hybrid approach can also be considered."

SOUTH AFRICA ENERGY CRISIS

A study examining power sector reorganizations, which included 142 developing countries from 1982 to 2013, shows that privatization and liberalization of competition continue to lag behind. Like South Africa, these countries are highlighted as having energy sector reforms that lack the full commitment to allowing IPPs and establishing electricity regulators to be fully integrated into the countries' energy sector [14]. This indicates that the current systems do not foster innovation in the energy sector. The study documents that the poorer the country, the lower the institutional capacity to support private power generation. Systems are crucial in implementation, as "hybrid power markets remain a reality, but economic growth and democratization can move power sectors in emerging economies in a different direction."

Transitioning to environmentally friendly energy requires a regime that supports private investment and complementary policies. Todd and McCauley [15] report that governments must bring together more foreign investment to initiate large-scale renewable energy projects, including repurposing current infrastructure to allow private power generators to supply energy to the national grid. This supports the idea that underdeveloped countries will face challenges in implementing and providing private power generation. Restructuring Eskom, the leading electricity producer in South Africa, is crucial given the country's ongoing crisis. Todd and McCauley [15] also note political pressures from mining and labor constituencies resisting Eskom's restructuring. Foreign expertise and diverse approaches may be critical in fostering a new commitment to energy transition in terms of finance and new energy systems.

Accurately determining electricity demand is essential for estimating current and future usage. A study by Ye et al. [16] finds that household income and electricity price are significant determinants of electricity demand in South Africa, which is normal and downward-sloping. The study notes that substitute fuel use impacts demand elasticity. Demand is higher for appliance-rich households in urban areas (in urban areas), especially with more household members and larger dwellings. South Africa's growing middle class, equipping their homes with modern appliances, suggests that electricity demand will increase despite calls to raise electricity tariffs. Thus, the government has a crucial role in this process.

Political will is crucial for supporting the transition and funding of projects that enable the shift to alternative energy sources. Todd and McCauley [15] emphasize the importance of the South African power system, noting that greater control over local power generation requires Eskom's cooperation in energy transfer through its transmission and distribution networks. This suggests that the transition to sustainable solutions will still depend on Eskom's efficiency and reliable infrastructure. Municipalities should also take the lead in promoting net metering to encourage distributed renewable energy. For an efficient energy market, it may be ideal for IPPs to sell electricity to municipalities. Municipal involvement is essential in fostering support for the energy transition.

INDEPENDENT POWER PRODUCTION AS A SOLUTION TO POWER CRISIS

Nepal et al. [17] describe IPPs as private companies that own and operate electricity generation facilities, selling power to utilities, central government buyers, and end users based on their business models. IPPs are viewed as a solution to state electricity crises where public utilities cannot meet demand, enhancing electricity supply by feeding power back into the national grid. Countries need to establish a power trading model, for strong political will and favorable economic conditions, to attract private investment in power generation. IPPs introduce competition, thereby reducing the monopoly power of public utilities.

AFRICAN LONG-TERM ENERGY SOLUTION

The study identified political risk, human capacity issues, and corruption as significant obstacles, along with governance gaps that stakeholders perceive as barriers to deploying renewable energy sources. These risks impede optimal market development for renewable energy and deter sustainable private investment. The research suggests that policy efforts should enhance the interaction between private and public partners by fostering greater consensus-building, transparency, stakeholder management, effective administration, and improved decision-making.

The South African Renewable Energy Independent Power Producer Procurement Programme (REIPPPP) is a competitive bidding initiative designed to attract private sector investment in grid-connected renewable energy generation. Since its inception in 2011, the REIPPPP has been highly successful, procuring 6328 MW of capacity from wind, solar photovoltaic, and other renewable sources and attracting USD 20.5 billion in investments. Notably, bid tariffs have significantly decreased, with recent projects achieving some of the world's lowest prices for grid-connected renewable energy. South Africa's success in attracting more investment through IPPs in four years than the rest of SSA has over the past 25 years underscored the REIPPPP's model as a valuable reference for policymakers in Africa and other developing regions [18].

Data from across the continent reveals diverse ownership and financing structures for IPPs, with development financing institutions (DFIs) playing a crucial role in risk mitigation and attracting private financiers. "The advancement of renewable energy in Africa, marked by both scale and cost reductions, is partially driven by competitive procurement or auctions, which offer lower prices and greater transparency compared to renewable energy feed-in tariffs or directly negotiated contracts" [19]. These developments emphasize the importance of dynamic, cost-effective planning, timely competitive procurement, strong regulatory capacity, and appropriate risk mitigation mechanisms.

Under national government guidance, South African provincial and local leaders have also crafted strategies to promote sustainable energy usage. The REIPPPP has enlisted over 92 power producers, contributing more than 6300 MW of power, predominantly from solar and wind sources. Evidence indicates that adopting green energy has lowered production costs, created jobs, attracted foreign investment, and garnered local stakeholder support. Many projects are located in rural areas, benefiting local communities through employment, development, and improved quality of life. Despite these achievements, challenges remain as South Africa continues to integrate this new industry, and there is still a significant need for investment in large-scale projects to address the country's ongoing challenges and further local economic development [20].

CORROBORATING CURRENT PLANS WITH AVAILABLE SOLUTION

South Africa's renewable energy revolution holds significant growth potential, driven by technological advancements, emerging energy markets, and favorable geographic conditions. The country's shift toward renewable energy is propelled by internal needs, such as diversifying its energy mix and reducing reliance on fossil fuels, as well as external pressures, including global commitments to combat climate change.

Advancements in renewable energy technologies have growth potential for Africa's energy sector. Innovations in solar photovoltaic (PV) panels, wind turbines, and battery storage systems have made renewable energy more efficient, reliable, and cost-effective. The declining costs of solar PV panels and wind turbines have made these technologies increasingly competitive with traditional fossil fuel-based power generation. As discussed earlier in the literature, with proper policies, these advancements address South Africa's energy crisis, improve living standards, and support the local economy. Zarfl and Peters [21] argued that 70% of Africa's energy could come from renewable sources, with 82% of this being hydropower, 11% solar power, and 7% wind power. In 2015, Nel et al. [22] from the University of Pretoria conducted a research on the risks and barriers to renewable energy development in South Africa, specifically focusing on independent power production.

Countries in situations similar to South Africa need a hybrid approach that ensures current energy sources are sufficient, opens the grid to IPPs, and gradually integrates renewable energy. Borrowing heavily to invest in green energy and increasing debt is not beneficial.

The emergence of new energy markets, both domestically and globally, offers additional opportunities for the growth of renewable energy in South Africa. As the global shift toward clean energy accelerates, demand for renewable energy technologies and expertise increases. South Africa's established renewable energy industry and its strategic location as a gateway to the African continent position it as a key player in the growing clean energy market.

The South African government's commitment to expanding renewable energy, as outlined in the REIPPPP, has attracted significant investment from domestic and international stakeholders. This program has facilitated the procurement of renewable energy projects through competitive bidding processes, driving down costs and expanding the country's renewable energy capacity.

EMPOWERING INDEPENDENT POWER PRODUCERS

The crisis of electricity in South Africa and the rest of Africa has deterred economic progress and contributed to lower economic growth in affected areas. Electricity is an essential input in production and a basic daily human requirement. Countries like South Africa have invested in ensuring that every household and business has access to electricity without improving its generation capacity to match the demand. The country is in a state that requires additional power generation; thus, it has allowed IPPs to be part of the solution to the energy crisis. To avoid having load shedding as a regular occurrence in the country, South Africa is also transitioning into a just energy strategy that aligns with the Paris Agreement on climate change. The last ten years have been a testimony that South Africa requires sustainable power to maintain positive economic growth. The literature in this study supports that the transition to environmentally friendly energy requires investments and complementary infrastructure and policy.

Based on the above, if the IPPs are to be a solution that is inclusive and open to all innovative initiatives, the government has to task development banks to support entrepreneurs who intend to get into power generation. This will enable a faster energy transition, supporting entrepreneurs' initiatives.

Involving IPP is a considerable solution in countries with corrupt officials; this takes some essential activities from the state to the private sector. In addition, it reduces the overall debt that may be required to finance the implementation of energy solutions.

ENERGY SOLUTION FOR THE AFRICAN CONTINENT

Renewable energy alone is not a comprehensive solution to the electricity crisis in Africa. Instead, African countries should consider a hybrid model that aligns with their development plans while addressing urgent environmental concerns. The literature indicates that alternative energy sources, though essential, come with significant costs and may conflict with some countries' developmental agendas. However, there is an urgent need to protect the planet from the adverse effects of climate change. Therefore, the solution lies in reducing reliance on fossil fuels and gradually transitioning to green energy rather than ultimately shifting investment solely to green energy.

HYBRID ENERGY MODEL WHICH BALANCES DEVELOPMENT AND SUSTAINABILITY

The president of South Africa has been vocal in calling for the diversification of energy sources to reduce the impact of load shedding in the country. This is because a hybrid energy model incorporates a mix of renewable energy sources, such as solar, wind, hydro, and geothermal, with traditional fossil fuels. This ensures a more stable and reliable energy supply while leveraging the strengths of each energy source. In South Africa, the government has started buying power from IPPs; it is electrified by a combination of solar and wind energy with natural gas, which can provide a steady power supply; in the long term, natural gas and coal can be used as backup when renewable sources are intermittent.

A gradual transition to green energy requires significant infrastructure investment, which is crucial to support both traditional and renewable energy sources. This includes upgrading transmission and distribution networks to handle the variability of renewable energy. Therefore, a phased approach to green energy adoption instead of an abrupt shift allows for a smoother transition. This includes setting realistic milestones for renewable energy integration while maintaining existing infrastructure. Also, investing in modernizing fossil fuel plants to reduce emissions can serve as an interim measure while renewable infrastructure is being developed.

South Africa is exemplary in developing a supportive policy and regulatory framework, which is essential to attract investment in both traditional and renewable energy sectors. Policies should encourage private sector participation, streamline approval processes, and provide incentives for renewable energy projects. Establishing clear regulations for IPPs and ensuring transparency in the procurement process can boost investor confidence.

Sustainable financing is vital for implementing a hybrid energy model. This includes leveraging international climate funds, public-private partnerships, and DFIs to mobilize capital. So far, many African countries need to transition from traditional energy sources to green initiatives because the resources to finance research and development (R&D) and green initiatives are limited, as shown in the literature. Innovative financing mechanisms such as green bonds can also attract investment in renewable energy projects.

Research and Development: R&D investment can drive innovation in traditional and renewable energy technologies. This includes exploring new ways to enhance the efficiency of renewable energy sources and developing cost-effective methods for carbon capture and storage (CCS) in fossil fuel plants. This progress is starting to be visible in South Africa, and collaboration with international research institutions can facilitate the transfer of knowledge and technology to African countries.

Since African countries are rich in natural resources, a hybrid energy model that balances renewable energy with traditional fossil fuels offers a practical pathway for African countries to address their electricity crises without compromising their development plans.

CONCLUSION AND RECOMMENDATION

Africa needs to involve its citizens in power generation initiatives, allowing those who generate surplus electricity to sell it to others in need. Opening the energy market to the general population will encourage innovation, foster entrepreneurship, reduce dependency on the government, and provide opportunities for income generation. For example, in China, rural areas have been mobilized to generate energy and income simultaneously. Specialized training, infrastructure, and supportive policies are necessary to enable the participation of the rural regions. This approach is highly recommended for Africa, which has significant potential to harness clean energy from the sun. China demonstrates that a country can only partially depend on fossil fuels, regardless of their abundance.

The authors cited in this study emphasize the importance of green energy and the need to protect the environment. The discussion highlights that Africa is uniquely positioned to consider alternative energy sources due to its geographic advantages. Simultaneously, Africa has natural resources such as coal and gas that can be used to address its energy crisis while seeking a permanent, sustainable solution. However, climate change is a global issue requiring urgent action. A hybrid approach is recommended, as it supports developmental policies. Involving IPPs in the energy transition is a viable method that does not require high state debt to finance the transition. The government's role is to provide complementary infrastructure to support energy producers.

REFERENCES

1. Jesson, J. K., & Lacey, F. M. (2006). How to do (or not to do) a critical literature review. *Pharmacy Education, 6*, 139–148.
2. Khan, K. S., Kunz, R., Kleijnen, J., & Antes, G. (2003). Five steps to conducting a systematic review. *Journal of the Royal Society of Medicine, 96*, 118–121.
3. Berahab, R. (2022). The energy crisis of 2021 and its implications for Africa. *Policy.* https://www.policycenter.ma/sites/default/files/PB_06-22_Berrahab.pdf
4. Analytics, C. (2022). Global reaction to energy crisis risks zero carbon transition. https://climateactiontracker.org/publications/global-reaction-to-energy-crisis-risks-zero-carbon-transition/ (accessed 2 July 2022).
5. Stiewe, C., Ruhnau, O., & Hirth, L. (2022). European industry responds to high energy prices: The case of German ammonia production. https://www.econstor.eu/bitstream/10419/253251/1/Stiewe%20et%20al.%202022.%20European%20industry%20responds%20to%20high%20energy%20prices.%20Econstor.pdf
6. Zheng, Y. H., Li, Z. F., Feng, S. F., Lucas, M., Wu, G. L., Li, Y., Li, C. H., & Jiang, G. M. (2010). Biomass energy utilization in rural areas may contribute to alleviating energy crisis and global warming: A case study in a typical agro-village of Shandong, China. *Renewable and Sustainable Energy Reviews, 14*, 3132–3139.

7. Hua, Y., Dong, F., & Goodman, J. (2021). How to leverage the role of social capital in pro-environmental behavior: A case study of residents' express waste recycling behavior in China. *Journal of Cleaner Production, 280,* 124376.
8. Pan, Y., & Dong, F. (2022). Dynamic evolution and driving factors of new energy development: Fresh evidence from China. *Technological Forecasting and Social Change, 176,* 121475.
9. Bogodukhova, E. S., Bobrova, E. O., Britvina, V. V., Zueva, A. S., & Sorokin, A. Y. (2021). Trends in forming an alternative energy balance in Russia using information technologies during the climate crisis. *IOP Conference Series: Earth and Environmental Science, 808,* 012009.
10. Islam, M. S., Al-Amin, A. Q., & Sarkar, M. S. K. (2021). Energy crisis in Bangladesh: Challenges, progress, and prospects for alternative energy resources. *Utilities Policy, 71,* 101221.
11. Coyle, E. D., & Simmons, R. A. (2014). *Understanding the Global Energy Crisis.* Purdue University Press.
12. Olatunji, O., Akinlabi, S., Oluseyi, A., Abioye, A., Ishola, F., Peter, M., & Madushele, N. (2018). Electric power crisis in Nigeria: A strategic call for a change of focus to renewable sources. *IOP Conference Series: Materials Science and Engineering, 413,* 012053.
13. Musonye, X. S., Davíðsdóttir, B., Kristjánsson, R., Ásgeirsson, E. I., & Stefánsson, H. (2020). Integrated energy systems' modeling studies for sub-Saharan Africa: A scoping review. *Renewable and Sustainable Energy Reviews, 128,* 109915.
14. Urpelainen, J., & Yang, J. (2019). Global patterns of power sector reform, 1982–2013. *Energy Strategy Reviews, 23,* 152–162.
15. Todd, I., & McCauley, D. (2021). Assessing policy barriers to the energy transition in South Africa. *Energy Policy, 158,* 112529.
16. Ye, Y., Koch, S. F., & Zhang, J. (2018). Determinants of household electricity consumption in South Africa. *Energy Economics, 75,* 120–133.
17. Nepal, R., Sofe, R., Jamasb, T., & Ramiah, V. (2022). Independent power producers and deregulation in an island based small electricity system: The case of Papua New Guinea. *Energy Policy, 172,* 113291
18. Eberhard, A., Gratwick, K., Morella, E., & Antmann, P. (2017). Independent power projects in sub-Saharan Africa: Investment trends and policy lessons. *Energy Policy, 108,* 390–424.
19. Eberhard, A., & Naude, R. (2016). The South African renewable energy independent power producer procurement programme: A review and lessons learned. *Journal of Energy in Southern Africa, 27,* 1–14.
20. Jain, S., & Jain, P. (2017). The rise of renewable energy implementation in South Africa. *Energy Procedia, 143,* 721–726.
21. Zarfl, C., & Peters, R. (2024) 76% of Africa's energy could come from renewable sources by 2040: Here's how the conversation. https://theconversation.com/76-of-africas-energy-could-come-from-renewable-sources-by-2040-heres-how-224699 (accessed 28 May 2024).
22. Nel, D. (2015). Risks and barriers in renewable energy development in South Africa through independent power production. *African Journal of Public Affairs,* 1–20.

Chapter 10

DQ current control strategies for single-phase grid-connected inverter

Pranati Rani Purohit, Arnab Ghosh, and Pravat Kumar Ray

INTRODUCTION

Sustainable energy reserves, such as solar photovoltaics and wind generators, generate electrical power in the direct current (DC) form. Inverters' utilization becomes essential to ease the conversion of direct (DC) electricity to alternating (AC) electricity. The utility of alternating current (AC) electricity is prevalent in residential households and the power grid. The successful implementation of this conversion is of utmost importance to guarantee compatibility with the current electrical infrastructure. Additional functions were needed to synchronize the power-driven intervening distribution network and small-scale embedded generators (DERs) due to the rapid expansion of power distribution systems and the networks connected to renewable energy sources (RESs). Safe and diverse energy options are provided by distributed generating systems (DGs), which also improve transmission and distribution efficiency, aid in reducing emissions of greenhouse gases, increase stability and power quality, and decrease energy prices. However, DGs rely on RESs, which aren't reliable energy sources. Power electronic inverters link distributed generation (DG) and RESs to the power grid. Due to their low current harmonics and high-power quality in bidirectional energy flow, DGs often employ single-phase inverters operating in grid-tied mode. A single-phase inverter has two modes of operation: islanding, or working alone, and grid-tied, or both. Depending on the setup, the system can send real or reactive power (P/Q) to the load in grid-tied mode.

Similarly, while operating in standalone mode, the grid can generate its power without relying on other grids for this function. The inverter's ability to distribute power during grid faults results from its controlling procedures. Shared real and reactive power in single-phase/three-phase systems where the grid is operated in standalone or grid-connected mode has been the subject of numerous research proposals. DGs rely on RESs, which aren't reliable energy sources. Power electronic inverters link distributed generation (DG) to the utility grid. Photovoltaic (PV)-based voltage source inverters (VSIs) govern the injected power into the grid using the proportional-integral (PI) controller. However, system unpredictability is caused

DOI: 10.1201/9781003628279-10

by the greater THD. For PV VSI, the system dynamics are simplified and improved in the current research articles. However, there are serious stability issues with the output power. It appears that the controllers being used have a delayed response time.

Effective DC-link voltage control is crucial for maintaining the photovoltaic (PV) system's maximum power point voltage. The effectiveness of this control directly impacts the quality of power transferred to the utility grid [1]. Simultaneously, the primary goals of the inverter operated in grid-connected mode are to achieve a decoupled P/Q control that is highly efficient (i.e., has a tiny tracking THD), improves energy quality, and complies with the Algerian Grid Code (AGC) [2].

This research suggests a method for controlling the real and wattless power components using a PI controller for a single-phase grid-tied inverter based on the literature findings. The rotating reference frame (RRF) is now employed for single-phase inverters according to direct quadrature (DQ) axis theory [3], which was initially developed for 3-phase inverters. One benefit in the context of this principle is that a PI controller can be used to control the AC variables, essentially the same as the DC quantities changed with primary frequency using RRF. In RRF, the PI controllers regulate the inverter's instantaneous power and the conventional current control system's output power [4]. The $\alpha\beta$ frame in DQ theory is analogous to the PI controllers. Because of its DQ transformation features, the DQ theory is also very successful when harmonic distortion is present [5]. The inverter's output is both harmonic and devoid of delays. The harmonics are reduced by designing and using an LCL filter [6]. To summarize the novel contributions, an inverter connected to single phase in the grid-tied mode with DQ current control driven by the PI control is suggested for real and wattless power regulation. Also, to lower the harmonics, an LCL filter is used [7,8].

Further, there are efficiency and cost concerns with traditional PV-based inverter systems that rely on transformers for isolation. Consequently, transformer-less structures are the more appropriate solution. Therefore, the current research is on the transformer-less structure.

SYSTEM MODELING AND DESIGN

The presented prototype of a DC-link voltage connected to a grid-tied inverter connected to a single phase is illustrated in Figure 10.1. Because of its simplicity and capacity to handle bidirectional power flow, single-phase full-bridge inverters are essential to power electronics theory and practice. Injecting energy into the grid is made possible by this topology's conversion of DC energy from DERs into AC energy. The study shows that power regulation's real and wattless component is made possible by organizing single-phase full-bridge inverter switching signals [9].

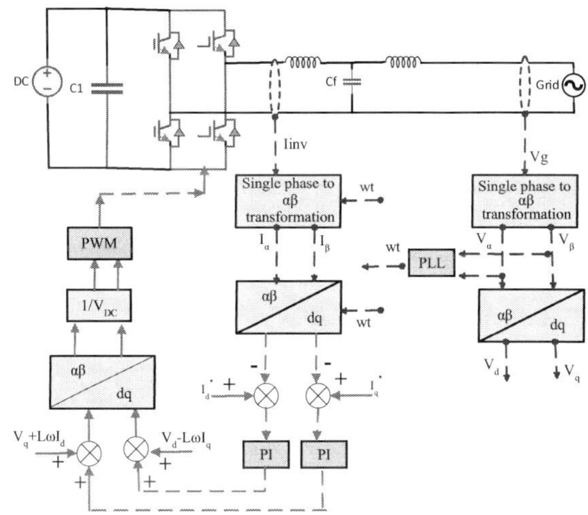

Figure 10.1 Current control architecture of inverter coupled to the grid in single phase.

Further, the inverter's bidirectional operation modes involve the current flow from the grid side toward the inverter and vice versa. Due to the pulse width modulation (PWM) approach involved in the inverter's output current control, it includes higher-order harmonics. For this reason, the L filter in the grid-to-inverter connection is often made. Nevertheless, harmonics that are already there are muted by the L filter. The L filter is expensive for high-power uses and has a sluggish system dynamic response. Also, the L filter causes a significant voltage loss.

The voltage and output current are converted into a DQ reference frame to improve the tracking performance of the suggested PI current control. In addition, the converted amounts are employed in both an interior and exterior loop to obtain the necessary signals. To generate the variable ωt, a phase lock loop (PLL) is implemented. The inverter's gate signals are produced through sinusoidal pulse width modulation (SPWM).

LCL Filter: The inverter's output currents have been corrected using a third-order LCL filter rather than the perpetual L filter [10]. Although the LCL filter has a resonance issue, it is known to be more successful than the L filter in decreasing switching harmonics. Moreover, the specification procedure for determining the parameter values of the LCL filter is significantly more intricate than that of the L filter. Selecting proper settings is crucial due to several restrictions, including filter size, current harmonics, and power absorption in the filter capacitor. Therefore, in this work, the inverter is connected to the grid via an LCL filter to bring down the current and voltage harmonics, as depicted in Figure 10.2. The upper limit of allowable ripple current flowing through the inverter is contingent upon the value of L_i, which represents the inductor on the inverter side responsible

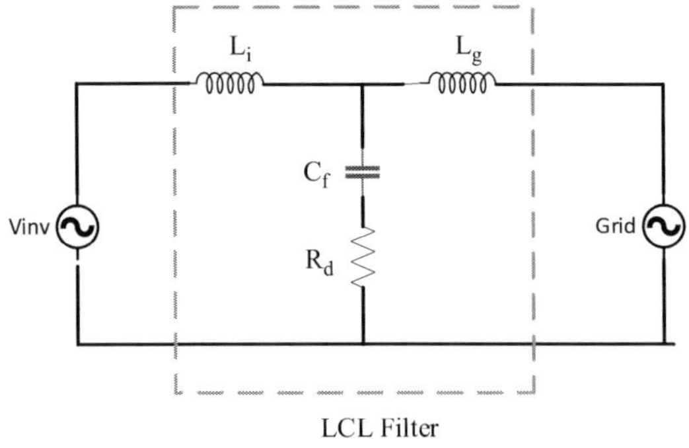

Figure 10.2 LCL filter structure.

for transporting current I. The inverter side inductance, L_i, grid side inductance, L_g, filter capacitance, C_f, and the damping resistance R_d can be determined from [11].

On the inverter side, the measured inductor current ripple is restricted to 20%, referred to as condition (10.1).

$$L_i = \frac{V_{DC}}{4 f_{sw} \Delta I_{ppmax}} \tag{10.1}$$

The selection of the total inductance ($L_1 + L_2$) is determined by the maximum voltage decrease surpassing the inductor. A maximum reduction of 10% in the measured voltage is allowed, referred to as condition (10.2).

$$V_{L_i+L_g} = I * 2\pi f \left(L_i + L_g\right) \tag{10.2}$$

The resonant frequency is represented as

$$\omega_{res} = \sqrt{\frac{L_i + L_g}{L_i L_g C_f}} \tag{10.3}$$

$$f_{res} = \frac{w_{res}}{2\pi} \tag{10.4}$$

The design of the capacitor is determined by the absorbed reactive power under the specified rated conditions. As an accepted guideline, capacitors can only consume reactive power up to 5% of their rated power.

The value of (S) is specified in equations (10.5) and (10.6), respectively.

$$Q = V^2 2\pi f \tag{10.5}$$

$$C_f = \frac{5\%S}{V^2 2\pi f} \tag{10.6}$$

It is recommended that a series-connected resistor be incorporated alongside the capacitor to mitigate the inconstancy and unsteady condition of the filter. This particular solution is occasionally referred to as 'passive damping.' Although straightforward and dependable, the system experiences increased heat dissipation and drastically diminishes the filter's efficiency. The formula to evaluate the damping resistor is as follows:

$$R_{\text{fd}} = \frac{1}{3\omega_{\text{res}} C_f} \tag{10.7}$$

Phase-Locked Loop: For an active current to be transmitted from the inverter to the utility grid, the voltage and current phases must lie in phase. Harmonizing the grid voltage with the current of the inverter fastened to the grid is essential for controlling the grid's real and wattless component of power [12]. The amplitude of the signal and the phase angle of the grid voltage are determined by the PLL technique. A grid-tied inverter's current control relies on this method, which produces a signal that acts as a reference. PLL technology creates a signal in phase with the measured voltage 90° out of phase to provide reactive power to the grid. A change from $\alpha\beta$ to dq frame is necessary for the control method. But because there aren't enough phase variables in single-phase systems, this transformation can't happen. Two orthogonal variables are required for this transformation in such situations. Therefore, to make the current control approach discussed earlier work for single-phase systems, we must build a fictional part perpendicular to the real one. Combining the real and imaginary signals in a stationary or β frame can be accomplished by shifting the measured actual signal in phase. A low-pass filter, which is applied with a voltage and has the given transfer function, will generate an alpha-beta voltage in a single-phase system.

$$\text{TF} = \frac{\omega_c}{s + \omega_c} \tag{10.8}$$

where α and β components of current values of PLL are denoted as

$$i_\alpha = i\cos(\theta_{\text{PLL}}) \tag{10.9}$$

$$i_\beta = \alpha \; i\sin(\theta_{\text{PLL}}) \tag{10.10}$$

Then, DQ transformation is applied to the α and β components of the current to derive the current's d and q components in the following way:

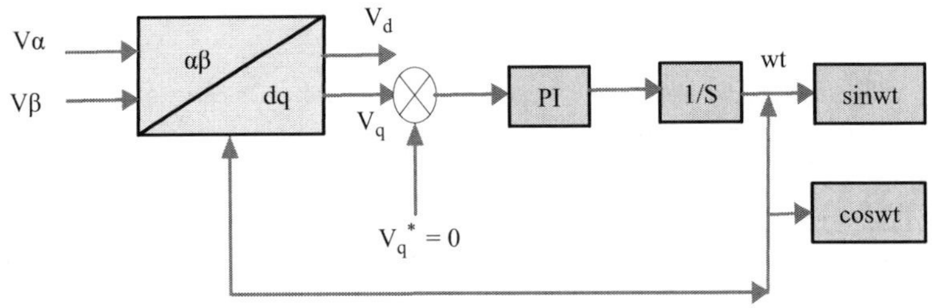

Figure 10.3 Phase-locked loop.

$$i_d = i(\cos^2 \theta_{\text{PLL}} + \sin^2 \theta_{\text{PLL}}) \tag{10.11}$$

$$i_q = i(-\cos\theta_{\text{PLL}} + \sin\theta_{\text{PLL}} + \cos\theta_{\text{PLL}} + \sin\theta_{\text{PLL}}) \tag{10.12}$$

By utilizing a PI controller, the $\cos(\omega t)$ and $\sin(\omega t)$ components are found, as shown in Figure 10.3.

The formula for calculating the inverter's real and wattless power component concerning the grid is denoted as follows: V_{inv} and V_g depict the inverter and grid voltage, respectively.

$$P = \frac{V_{\text{inv}} V_g}{\omega L} \sin\delta \tag{10.13}$$

$$Q = \frac{(V_{\text{inv}} - V_g)V_g}{\omega L} \tag{10.14}$$

METHODOLOGY OF CONTROLLING CURRENT

The controller design is the basis upon which the whole control diagram is constructed, as illustrated in the literature. Regarding controller processes, PI are recommended as the most popular mechanism for ensuring actions with multiple functions.

The interior current regulation loop and the exterior voltage regulation loop are set to operate on PI controllers. Power quality, total harmonic distortion (THD), and grid current control are all under the purview of the inner loop of current control. In contrast, the exterior loop governs active and reactive power flows. Further, PWM signal regulation is the responsibility of the PI controller. A higher-quality current waveform can be generated by adjusting the PWM, which regulates the necessary output current from the inverter [13]. A sine wave and a carrier wave are included in the control transfer function. Modifying the amplitude and phase of a signal enables one to acquire its output voltage.

One of the outer loops in DQ frameworks regulating active power is the beta input signal, a d-component [14]. On the other hand, the alpha output signal controls the system's reactive power and is thus a q-component.

Zero steady-state error and the usage of DC quantities are only two of the many great qualities of the current controllers that use DQ-frame components, which give them a broad range of potential applications [15]. The control strategy proposed in [16] efficiently eliminates disturbances from the source and maintains low current distortion values, even when the grid voltage is severely distorted. It is based on PI controllers using grid-following inverters that operate on a single phase as part of their selective active, reactive, and distortion (PQD) power control strategy. Also, the simplicity of a PI controller makes it a popular choice for DC-link voltage tracking [17].

Based on the DQ reference frame theory, the components of real and wattless current within the DQ-frame are denoted in the following manner.

$$i_d = A\sin(\omega t + \delta) \tag{10.15}$$

$$i_q = A\sin(\omega t + \delta - \pi/2) \tag{10.16}$$

$$\begin{vmatrix} i_d \\ i_q \end{vmatrix} = \begin{bmatrix} \cos(\omega t) & \sin(\omega t) \\ -\sin(\omega t) & \cos(\omega t) \end{bmatrix} \begin{vmatrix} i_\alpha \\ i_\beta \end{vmatrix} \tag{10.17}$$

$$i_d = A\sin(\omega t + \delta)(\sin\omega t)\ A\sin(\omega t + \delta - \pi/2)(-\cos\omega t) \tag{10.18}$$

$$i_q = A\sin(\omega t + \delta)(\cos\omega t)\ A\sin(\omega t + \delta - \pi/2)(\sin\omega t) \tag{10.19}$$

$$\begin{vmatrix} V_d \\ V_q \end{vmatrix} = \begin{bmatrix} \cos(\omega t) & \sin(\omega t) \\ -\sin(\omega t) & \cos(wt) \end{bmatrix} \begin{vmatrix} V_\alpha \\ V_\beta \end{vmatrix} \tag{10.20}$$

$$V_{d1}^{*} = V_{d1} + \omega L i_q - V_d \tag{10.21}$$

$$V_{q1}^{*} = V_{q1} - \omega L i_d - V_q \tag{10.22}$$

V_d and V_q represent the grid voltage's d and q axes, respectively. The components of the output grid voltage on the d and q axes, respectively, are V_{d1} and V_{q1}. The symbol ω represents the grid angular frequency.

PERFORMANCE EVALUATION

An inverter connected to the utility grid in single phase can benefit from the suggested control scheme, according to MATLAB/Simulink results. This section already discussed the use of the DQ-frame transformation. The Simulink model's design and implementation are completed within a time frame of 0.2 s using MATLAB. The current values are defined concerning the 10-A reference currents i_d and i_q. The controller's step reaction time is relatively high, and the present signal response values are adjustable with reference values. All of the system's electrical Simulink model parameter values are displayed in Table 10.1. It is necessary to calibrate the gains of PI control so that the execution always aligns with the desired response. The output voltage of the grid is 380 V rms (root mean square), as represented in Figure 10.4. Figure 10.5 displays the output inverter current. Figure 10.6 shows the PLL output. The total harmonic value was evaluated to be less than 5%, as shown in Figure 10.7.

Table 10.1 Parameters of simulation

Symbol	*Description*	*Values*
V_{DC}	DC voltage source	400 V
V_g	Grid voltage	325 V
f	Grid frequency	50 Hz
C_{DC}	DC bus capacitance	500 microF
P	Rated power	2 kVA
f_{sw}	Switching frequency	10 kHz
L_{inv}	Filter Inductance of inverter side	4.06 mH
L_g	Filter Inductance of grid side	4.35 mH
C_f	Filter capacitance	6.01 microF
k_p	Proportional gain	40
k_i	Integral gain	6.7

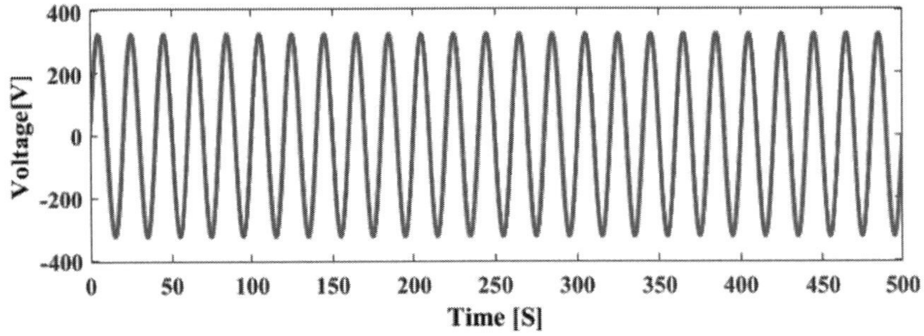

Figure 10.4 Output voltage.

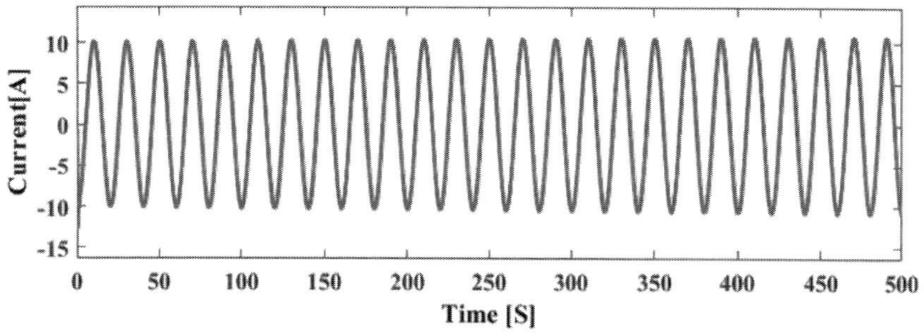

Figure 10.5 Output current.

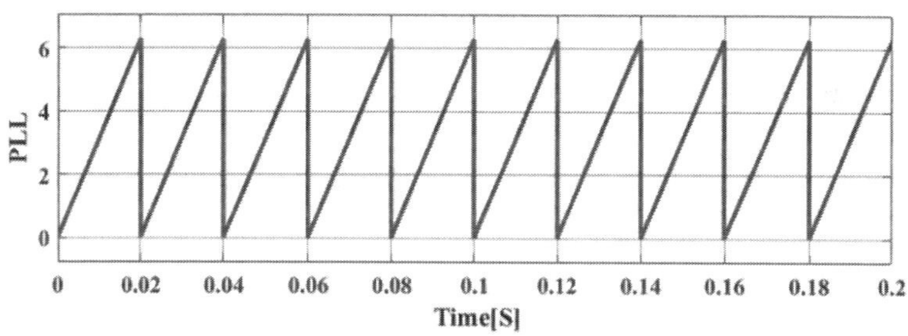

Figure 10.6 PLL output.

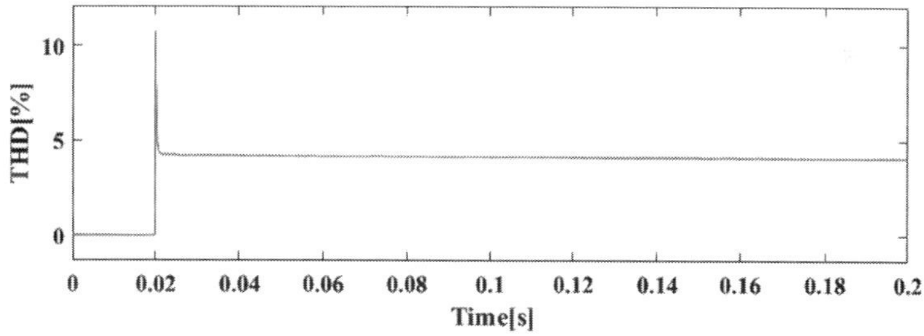

Figure 10.7 Total harmonic distortion [%].

CONCLUSION

A double-loop control scheme for a grid-following inverter linked to a DC source in a single-phase system is suggested in this chapter. This work gives the current control mechanism of an inverter operated in grid-connected mode in single-phase systems based on the DQ reference frame – a PI-based control approach with two closed loops guaranteed further minimal DC-link voltage fluctuation. Also, the technique employed manages the active and reactive power. The inverter's output is protected against harmonic distortion with the help of an LCL filter. The simulation results obtained from MATLAB Simulink are displayed here. The load's active and reactive power needs are met by regulating the inverter output. Per the IEEE standard requirement, the results conclude that the inverter's power transmission to the grid is under control, and the THD value is below 5%. The simulated output current and voltage values demonstrate the efficacy of the recommended PI controller in conjunction with the utility grid. Coordinated control and PV applications are made possible by the suggested control, which follows power and current references and performs well in steady-state with low distortion values. In addition, the findings from simulations show that the convergence is faster, and the steady-state voltage and current tracking accuracy are excellent.

REFERENCES

1. Inverters GPV, Aamri FEL, Maker H et al. Stability analysis for DC-link voltage controller design in single-stage single-phase. *IEEE J Photovoltaics*. 2023;13(4):580–589. doi:10.1109/JPHOTOV.2023.3263253
2. Beddar A, Abadlia I, Abdoune F, Hassaine L, Bengourina MR. Investigation and implementation of a nonlinear decoupled power control for a single-phase grid-connected power inverter via an LCL filter. *J Eng Res*. 2023;11(1):100010. doi:10.1016/j.jer.2023.100010
3. Tang X, Chen W, Zhang M. A current decoupling control scheme for LCL-type single-phase grid-connected converter. *IEEE Access*. 2020;8:37756–37765. doi:10.1109/ACCESS.2020.2976110
4. Shi J, Huang W, Zheng X. Complete decoupling compensation of three-phase inverter with LCL filter in synchronous reference frame. *IET Power Electron*. 2022;15(5):381–394. doi:10.1049/pel2.12237
5. Zhang H, Xian J, Shi J, Wu S, Ma Z. High performance decoupling current control by linear extended state observer for three-phase grid-connected inverter with an LCL filter. *IEEE Access*. 2020;8:13119–13127. doi:10.1109/ACCESS.2020.2965650
6. Peña-Alzola R, Liserre M, Blaabjerg F, Ordonez M, Yang Y. LCL-filter design for robust active damping in grid-connected converters. *IEEE Trans Ind Informatics*. 2014;10(4):2192–2203. doi:10.1109/TII.2014.2361604
7. Khan D, Hu P, Habib S, Waseem M, Lin Z, Ahmed EM. A resonant damping control and analysis for LCL-type grid-connected inverter. *Energy Rep*. 2022;8:911–928. doi:10.1016/j.egyr.2021.11.273
8. Sangani SH. A Comprehensive analysis of a digital control strategy for photovoltaic-based single-phase grid-tied inverter systems. 32nd International Conference on Electrical Engineering (ICEE). 2024: 1–7.
9. Yilmaz U, Turksoy O. ScienceDirect Artificial intelligence based active and reactive power control method for single-phase grid connected hydrogen fuel cell systems. *Int J Hydrogen Energy*. 2022;48(21):7866–7883. doi:10.1016/j.ijhydene.2022.11.211

10. Saïd-Romdhane M, Ben Naouar MW, Belkhodja IS, Monmasson E. Simple and systematic LCL filter design for three-phase grid-connected power converters. *Math Comput Simul.* 2016;130:181–193. doi:10.1016/j.matcom.2015.09.011
11. Peña JCU, Sampaio LP, de Brito MAG, Canesin CA. RLC passive damped LCL single-phase voltage source inverter with capability to operate in grid-connected and islanded modes: design and control strategy. *Electr Eng.* 2020;102(4):2509–2519. doi:10.1007/s00202-020-01045-z
12. Xie Z, Chen Y, Wu W et al. Modeling and control parameters design for grid-connected inverter system considering the effect of PLL and grid impedance. *IEEE Access.* 2020;8:40474–40484. doi:10.1109/ACCESS.2019.2950933
13. Ghasemi MA, Zarei SF, Peyghami S, Blaabjerg F. A theoretical concept of decoupled current control scheme for grid-connected inverter with l-c-l filter. *Appl Sci.* 2021;11(14). doi:10.3390/app11146256
14. Dong N, Yang H, Han J, Zhao R. Modeling and parameter design of voltage-controlled inverters based on discrete control. *Energies.* 2018;11(8). doi:10.3390/en11082154
15. Inverters SG, Tedeschi E, Member S. Selective PQD Power Control Strategy for Single-Phase Grid-Following Inverters. *IEEE Journal of Emerging and Selected Topics in Power Electronics.* 2024;12(2):1294–1302. doi:10.1109/JESTPE.2023.3263796
16. Coteli R, Kayisli K, Caglayan RZ. Robust active and reactive power control by using single-phase grid connected H-bridge inverter. In: 2024 *12th Int Conf Smart Grid.* 2024:614–617. doi:10.1109/icSmartGrid61824.2024.10578081
17. Jain RK, Barry VR, Gadiraju HKV. An effective control strategy for single-phase single-stage PV grid-tied inverter under abnormal grid conditions. *IEEE J Emerg Sel Top Power Electron.* 2024;12(2):1249–1260. doi:10.1109/JESTPE.2023.3321374

Chapter 11

Chaotic quasi-oppositional differential search algorithm for transient stability constraint optimal power flow problem

Nirmalya Dey, Arko Ghosal, Sourav Paul, Sneha Sultana, and Provas Kumar Roy

INTRODUCTION

The term "power system" refers to a network that delivers energy from generators to loads. Energy from generators to loads is transferred via the electric power transmission network. Due to huge power transfers across longer distances, complex coordination, the difficult connection between numerous system controllers, and low power reserves, the current power system network is becoming more and more complex for planning and operating purposes. When unexpected interruptions can be tolerated with little loss of service quality, a power system is said to be secure. Before reaching an acceptable stable state where all operational constraints are kept within the allowed bounds, the system goes through the following transient states whenever there is a disruption. Within acceptable limits, the true and reactive powers of the generators in an interconnected power system should be flexible to use the least amount of fuel while still fulfilling the demands of the individual loads.

The planning, regulation, and operation of power systems are based on the idea of optimal power flow (OPF). The first to present it was Dommel and Tinney [1]. The community of power system researchers then put out other OPF strategies after that. At this point, OPF is a common power system analysis method and has shown to be reliable enough for real-world use. Thus, developing a power generation system that lowers fuel costs while satisfying all justice and inequality requirements is the main goal of the OPF solution. An OPF problem is usually a multi-objective, wide-ranging, highly nonlinear, non-convex optimization problem. The ideal steady-state operation of a power system must be established to minimize a desired objective function while still meeting the power balancing equations and any inequality limitations in the system. A nonlinear programming problem is used for this. The constraints on inequality are the limits of the independent and dependent variables, as well as the load flow equations, which are the equations for power balancing. Independent variables include the generator real powers, reactive power injection, transformer

DOI: 10.1201/9781003628279-11

tap settings, generator bus voltages, and omitting slack bus power. The dependent variables include generator reactive powers, line flows, slack bus power, and load bus voltages. Voltage instability is a new problem for power system construction and operation, even though the OPF's primary purpose is to reduce fuel costs. This is due to the ongoing rise in energy demand and the unmatched expansions in production and transmission capacity. Significant transmission loss may result from inadequate reactive power sources in electrical systems. It might be necessary in these circumstances to include transmission loss and voltage stability margin in the OPF issue's purpose. Lobato et al. [2] suggested an OPF based on linear programming (LP) by reducing losses in generator reactive margins and transmission losses in the Spanish power system. Using integer variables, this approach mimics the discontinuous properties of capacitors and shunt reactors. The approach that linearized the objective function and the restrictions in every iteration turned out to be better than one that merely linearized the objective function once. A generalized quadratic-based model for the OPF was used by Momoh [3] by expanding upon the fundamental Kuhn-Tucker constraints. The conditions for optimality, convergence, and feasibility were considered in the suggested OPF technique. It could also use hierarchical structures to incorporate different objective functions and selectable limitations. The generalized technique finds a globally optimal solution by combining optimal performance with the sensitivity of a target function to changes in constraints. Algorithmic memory and execution performance requirements were lowered. The Newton-based methods include the reduced gradient method, Lagrangian relaxation (LR), interior point methods (IPM) (Yan and Quintana [6]; Momoh et al. [7]), and LP (Pudjianto et al [8]; Momoh and Zhu [4]; Sun et al. [5]). Among the conventional approaches to OPF issue resolution that have developed over the past 20 years are those by Roy and Mandal [9] and Abou and Abido [10] that were found in the literature by Santos and Da Costa [11], Carvalho et al. [12], and Capitanescu and Wehenkel [13]. These methods find the overall optimal solution by using convexity and the starting point. However, complex objective functions that are not differentiable could not be resolved by techniques. One possible solution to the nonlinear objective function and constraint difficulties is to use nonlinear programming or NLP (Momoh et al. [7]). However, there are also several drawbacks to this approach, such as unpredictable convergence characteristics and an excessive number of numerical iterations that result in high processing costs, lengthy execution times, and complex algorithms. While a few of these techniques are efficient in identifying the optimal solution, they often converge slowly and incur significant computational costs. One drawback of Newton-based methods is that their merging characteristics depend on the beginning conditions; if the conditions are not right, they might not even converge. Lately, there have been notable gains made in power supply network optimization through heuristic approaches. To find the best solution or to be highly effective in addressing nonlinear OPF issues, many heuristic algorithms have been proposed recently. These include the following: genetic algorithm (GA) (Gaing and Chang [14]; Lai et al. [15]), simulated annealing (SA) (Roa-Sepulveda and Pavez-Lazo [16]), transition state (TS) (Abido [17]), particle swarm optimization (PSO) (Abido 17]; Swain and Morris [18]), chaotic ant swarm optimization (CASO) (Cai et al. [19]), bacterial foraging optimization (BFO) (Tripathy and Mishra [20]), differential evolution (DE) (Ela et al. [21]; Sivasubramani and Swarup [22]), biogeography-based optimization (BBO) (Roy et al. [23]), harmony search algorithm (HSA) (Khazali and Kalantar [24]), artificial bee colony (ABC) (Ayan and Kılıç [25]), and gravitational search algorithm (GSA) (Roy et al. [26]) are all examples of hybrid evolutionary programming (HEP). They have been useful in resolving several problems where it is better to find a global solution than a local one or when there are non-differentiable regions in the problem. Using FACTS instruments in IEEE 30-bus and IEEE 57-bus systems, Basu [27] applied the multidimensional discrete evolution approach to the OPF problem. The results were compared with earlier research. A non-dominated sorting gravitational search algorithm (NSGSA) was presented in 2012 by Hadi et al. [28] to update the gravitational acceleration. To evaluate the OPF issue, this (Sivasubramani and Swarup [22]) introduced the multi-objective HSA. In this study, the results of the proposed method were compared with non-dominated sorting genetic algorithm (NSGA-II), and several goals were taken into account, such as fuel consumption, L-index, and transmission loss. The overall fuel cost of thermal units, total emission, total real power losses, and voltage deviation were compared to other methods. A method for resolving the restricted optimal voltampere reactive (VAR) dispatch issue with objective functions that included fluctuating voltages and actual power interruptions was presented by Abido and Bakhashwain [29]. We evaluated the robustness and accuracy of the proposed method with the IEEE 30-bus system. The OPF with transient stability limits is one of the many OPF problems; it is a fascinating area for research because it considers all security and stability requirements in addition to some optimal solutions. Transient stability constraints are the limits that system dynamics must adhere to remain stable during a transient period. Transient instability causes very expensive synchronous loss in modern power systems. It is more efficient to use transient stability-constrained OPF to balance the power system's economics and security (OPF-TSC). Combining steady-state goals and stability constraints' economic objectives into a single, unique formulation is the primary purpose of the TSC-OPF. One method is to approach the TSC-OPF as an algebraic optimization problem, which may be solved by applying normal optimization techniques (Gan et al. [30]). After that, the TSC-OPF problem is understood by converting differential equations as inequality constraints into analogous algebraic equations. Based on this, the multi-contingency notion originated with Yuan et al. [31], who created a novel technique for adding a model of transient stability into traditional OPF that greatly reduces computational load. However, due to the large number of variables, it has serious limitations, including computation mistakes for approximations and convergence problems. The TSC-OPF problem has an infinitesimal dimension. To address this, Chen et al. [32] present a novel method termed functional transformation. However, these methods are sadly not relevant to real-world systems because of the high sensitivity issue of the initial conditions. The transient stability limit and the flexibility of the objective functions are two sorts of restrictions in OPF concerns that are constrained due to the computational challenges. Therefore, it is crucial to create optimization strategies that can overcome these constraints and handle these difficulties. Several population-based optimization strategies have been successfully applied to TSC-OPF problems, including the ABC by Karaboga et al. [33], the improved GA by Chan et al. [34], and differential evolution by Cai et al. [35]. This chapter presents the researcher's proposed chaotic quasi-oppositional-based differential search algorithm (CQODSA), which considers the Western Systems Co-ordinating Council (WSCC) 3-generator 9-bus system and the New England 10 generator 39 bus system to handle TSC-OPF difficulties.

FORMULATION OF MATHEMATICAL PROBLEMS

Objective function

OPF aims to minimize the objective function while maintaining all constraints related to equality and inequality. The following is one of the mathematical ways to phrase the OPF issue:

$$\text{Minimize}\,\text{OF}(x, y) \tag{11.1}$$

$$\text{Subject to}: \begin{Bmatrix} e(x, y) = 0 \\ ie_i \le ie(x, y) \le ie_u \end{Bmatrix} \tag{11.2}$$

where $\text{OF}(x, y)$ is the objective function; $e(x, y)$ are the equality constraints; $ie(x, y)$ are the inequality constraints; ie_i, ie_u are the minimum and maximum limits of the inequality constraints; x is the vector of dependent variables composed of slack bus active power, load voltages, reactive power of generators, and loadings of transmission lines; y is the vector of independent variables composed of reactive power injections, generator voltages, tap settings on transformers, and active power injections from generators other than the slack bus. Thus, the expression for x and y are

$$x = \left[P_{g1}, V_{l_1} \cdots V_{l_{\text{NL}}}, Q_{g_1} \cdots Q_{g_{\text{NG}}}, S_{l_1} \cdots S_{l_{\text{NTL}}} \right] \tag{11.3}$$

$$y = \left[P_{g_2} \cdots P_{g_{\text{NG}}}, V_{g_1} \cdots V_{g_{\text{NG}}}, T_1 \cdots T_{\text{NT}}, Q_{c_1} \cdots Q_{c_{\text{NC}}} \right] \tag{11.4}$$

where NG and NL are the number of generator and load buses; NTL is the transmission line's number; NT and NC are the number of transformers and shunt compensators.

The overall generation cost is taken into consideration as the optimal solution in this study. For the quadratic cost function, the total fuel cost of the generating units is given by:

$$F_c(u,v) = \sum_{p=1}^{\text{NG}} (a_p P_{g_p}^2 + b_p P_{g_p} + c_p) \tag{11.5}$$

where a_p, b_p, c_p are fuel cost coefficients, and P_{g_p} is the generation of the pth generator.

The valve point impact should be considered when presenting an acceptable cost function since many valve steam turbines are used in real generators to provide more precise and adaptive functioning. For generating units with valve point loading, the entire fuel cost is covered by:

$$F_c(u,v) = \sum_{p=1}^{\text{NG}} (a_p P_{g_p}^2 + b_p P_{g_p} + c_p + d_p \sin[e_p (P_{g_p}^{\min} - P_{g_p})]) \tag{11.6}$$

where d_p, e_p are the fuel cost coefficients with valve point loading effect and $P_{g_p}^{\min}$ is the minimum generation of the pth generator.

Transient stability assessment

To characterize the transient behavior of a system integrating transient stability, the generator rotor angles are provided here with respect to the inertial centers of each generator. The following location is the center of inertia, according to [36]:

$$\delta_{\text{COI}} = \frac{\sum_{k=1}^{N_{\text{NG}}} M_k \delta_k}{\sum_{k=1}^{N_{\text{NG}}} M_k} \tag{11.7}$$

where M_k is the inertia constant of the kth generator, δ_k is the rotor angle of the kth generator. The rotor angle constraint may be phrased as follows in terms of the inequality constraints:

$$\delta' \leq \delta_k - \delta_{\text{COI}} \leq \delta'' \tag{11.8}$$

$$K \subseteq S_G \tag{11.9}$$

where the set of generators is S_G and the lower and upper limits of generator rotor angles are δ', δ''.

Constraints

Equality constraints

The equality constraints that represent the load flow equations are as follows:

$$\begin{cases} P_{gi} - P_{Li} = \sum_{j=1}^{N} |V_i||V_j|\left(G_{ij}\cos\delta_{ij} + B_{ij}\sin\delta_{ij}\right) \\ . \\ Q_{gi} - Q_{Li} = \sum_{j=1}^{N} |V_i||V_j|\left(G_{ij}\sin\delta_{ij} - B_{ij}\cos\delta_{ij}\right) \end{cases} \tag{11.10}$$

where V_i, V_j are the voltages of the ith and the jth bus, respectively; P_{gi}, Q_{gi} are the active and reactive powers, respectively, of the ith generator; P_{Li}, Q_{Li} are active and reactive power, respectively, of the ith load bus; G_{ij}, B_{ij}, δ_{ij} are the conductance, admittance, and phase difference of voltages between the ith and the jth bus; N is the number of buses.

Inequality constraints

Generator constraints

The ith bus's generator voltage, reactive power, and active power are all between their minimum and maximum limits, as represented by:

$$\begin{cases} V_{gi}^{\min} \leq V_{gi} \leq V_{gi}^{\max} & i = 1, 2, \ldots. \text{ NG} \\ P_{gi}^{\min} \leq P_{gi} \leq P_{gi}^{\max} & i = 1, 2, \ldots. \text{ NG} \\ Q_{gi}^{\min} \leq Q_{gi} \leq Q_{gi}^{\max} & i = 1, 2, \ldots. \text{ NG} \end{cases} \tag{11.11}$$

where $V_{g_i}^{\min}$, $V_{g_i}^{\max}$ are the minimum and maximum generator voltages, respectively, of the *i*th generating unit; $P_{g_i}^{\min}$, $P_{g_i}^{\max}$ are the minimum and maximum active powers, respectively, of the *i*th generating unit; $Q_{g_i}^{\min}$, $Q_{g_i}^{\max}$ are the minimum and maximum reactive powers, respectively, of the *i*th generating unit.

Load bus constraints

$$V_{l_i}^{\min} \le V_{l_i} \le V_{l_i}^{\max} \qquad i = 1, 2, \ldots, \mathrm{NL} \tag{11.12}$$

where $V_{l_i}^{\min}$, $V_{l_i}^{\max}$ are the upper and lower load voltage range, respectively, of the *i*th generating unit.

Transmission cable limitations:

$$S_{li} \le S_{l_i}^{\max} \qquad i = 1, 2, \ldots \mathrm{NTL} \tag{11.13}$$

where S_{l_i} is the apparent power flow of the *i*th branch; $S_{l_i}^{\max}$ is the maximum apparent power flow limit of the *i*th branch.

Transformer bus constraints

Transformer tap settings are restricted to the following higher and lower limits:

$$\mathrm{TC}_i^{\min} \le \mathrm{TC}_i \le \mathrm{TC}_i^{\max} \qquad i = 1, 2, \ldots, \mathrm{NTL} \tag{11.14}$$

where $\mathrm{TC}_i^{\min}, \mathrm{TC}_i^{\max}$ are the minimum and maximum tap setting limits, respectively, of the *i*th transformer.

Constraints on shunt compensators

The following are the limits of shunt boundaries:

$$Q_{c_i}^{\min} \le Q_{c_i} \le Q_{c_i}^{\max} \quad i = 1, 2, \ldots, \mathrm{NL} \tag{11.15}$$

where $Q_{ci}^{\min}$, $Q_{ci}^{\max}$ are the minimum and maximum VAR injection limits, respectively, of the *i*th shunt capacitor.

DIFFERENTIAL SEARCH ALGORITHM (DSA)

Inspiration of DSA

The differential search algorithm (DSA) developed by Civicioglu is a revolutionary meta-heuristic optimization technique that is covered in this section [37]. DSA draws inspiration from the migratory patterns that organisms naturally follow. Owing to variations in seasons over the year, there are variations in the efficacy and accessibility of natural food sources every year. As a result, migration behavior is displayed by living creatures all year long. Because of this, organisms can move from one habitat to another, where resources are more abundant and can be used more efficiently. In the literature; this acts as the main motivation for calculating DSA [37].

Modeling of DSA mathematically

Initialization

In DSA, the initial population, referred to as the *artificial organism*, is created by a random process that highlights the challenge of optimization. Equation (11.16) [37] describes the initialization of a DSA synthetic organism.

$$X(i,j) = \text{rand} \times \left(ub(j) - lb(j)\right) + lb(j) \tag{11.16}$$

where $i=1, 2, \ldots.\ n_{p,}\ j=1, 2,\ldots, \dim$, $X(i,j)$ is depicted as the initial population of the jth variable of ith agent, *rand* is a number taken randomly in between $[0, 1], n_p$ and dim are depicted as the number of search agents (superorganism) and a number of control variables of the optimization problem, and $ub(j)$ and $ib(j)$ are the maximum and minimum limits of the jth variable, respectively.

Superorganism reshuffle at random

To locate the *stopoversite*, which is essential to an efficient process, select a *superorganism* at random from the starting population and move it in the direction of the target *donor* using random shuffling. Measure *scale* gamma random generation to control the functions of synthetic equation (11.17) [37].

$$\text{scale} = \text{randg}(2 \times r_1) \times (r_2 - r_3) \tag{11.17}$$

where randg is a gamma distribution value taken at random and r_1, r_2, and r_3 are numbers that are picked at random through $(0, 1)$. In DSA, the stopoversite of superorganism is presented in reference [37].

Boundary conditions

Two more variables, p_1 and p_2, are the maximum number in DSA. Members of that group were selected at random from the *superorganism* to interact at the *stopoversite*. Usually, values are generated in the range [0, 0.3]. A boundary condition is imposed on the *stopoversite* components by forcing them into the ecosystem once a portion of the *stopoversite* surpasses environmental limitations [37].

QUASI-OPPOSITIONAL-BASED LEARNING (Q-OBL)

OBL was initially proposed by Tizhoosh [38] as a ground-breaking idea for intelligence- or soft-virtualization strategies to advance various simulated processes. It seems to be one of the most effective theories in algorithmic wisdom for solving nonlinear optimization issues, improving the capacity of conventional population-based optimization techniques for searching the establishment. The OBL process starts with the initial estimate, which is random or predicated on some prior knowledge about the solution. In any case, or at the very least, the opposite may be the best course of action. To find a better solution along the path of optimal solutions, the opposing set of estimates is taken into consideration for convergence, constantly updating the initial estimates.

Opposite number

Let the real number be denoted by $P \in [y, z]$ and P^0 be its corresponding opposite number defined by:

$$P^0 = y + z - P \tag{11.18}$$

Opposite point

Say $P = (X_1, X_2, ..., X_n)$ is a point in n-dimensional space, where $P_r \in [y_r, z_r]$, $r \in \{1, 2, ..., n\}$. The opposite point $R^\circ = (X_1{}^\circ, X_2{}^\circ, ...X_n{}^\circ)$ is defined by its components:

$$P_r{}^\circ = y_r + z_r - P_r \tag{11.19}$$

Quasi-opposite number

Let P be a real number between $[y, z]$. The quasi-opposite number P^{Qo} is defined as

$$P^{Qo} = \text{rand}\left(C, \tilde{P}\right) \tag{11.20}$$

where C is given by: $C = \frac{y+z}{2}$.

Quasi-opposite point

Let P be a real number between $[a,b]$. The quasi-opposite point P_r^{Qo} is defined as

$$P_r^{Qo} = \text{rand}\left(C_r, \widetilde{P_r}\right) \tag{11.21}$$

where $C_r = \frac{y_r + z_r}{2}$, $P_r \in [y_r, z_r]; r = \{1,2,3,....,n\}$.

Chaotic DSA

To mitigate the drawbacks of DSA, chaotic DSA (or CDSA) was developed, fusing chaotic action with differential search optimization (DSA). The speed at which chaos, by its very nature random and non-repeating, conducts generic searches is crucial for accelerating the pace of convergence of a deify algorithm. There are ten chaotic maps in a chaotic set, and each one shows a distinct action. It has been established that the beginning point of this set is 0.7 between 0 and 1. The numerous chaotic map operations aid in resolving problems with convergence speed and local optimum.

CQODSA STEPS FOR OPF PROBLEM

Several stages that the CQODSA takes to solve the OPF problem are as follows:

- **Step 1:** Establish *superorganism* and randomly generate the first wave of individuals (P) for each person. Additionally, compute the feature subset in the designated subspace and construct the power system stabilizer (PSS) control parameters.
- **Step 2:** Create a quasi-opposing chaotic population and determine the fitness function.
- **Step 3:** Sort the group by fitness value, starting with the best members. *PE* stands for the best response obtained thus far. Each particle is assigned an initial *KE*.
- **Step 4:** Value dictates that privileged status-preserved solutions are limited to a small subset of the best answers.
- **Step 5:** To vary the independent variables, the solutions that do not belong to the elite group are again exposed to the varied chaotic map conduct of the CQODSA. The optimal site is thought to be where the fixates are because whales in CQODSA specifically seek the focuses.

Step 6: *Superorganisms* are shuffled at random, and the scale is assessed using the gamma distribution.

Step 7: The method is completed and the outcomes can be observed if the stopping requirement is satisfied; otherwise, proceed to **Step 4**.

SIMULATION RESULTS AND DISCUSSION

The applicability and validity of the various optimization algorithms were assessed using two test systems: the New England 10-generator 39-bus system and the WSCC 3-generator 9-bus system. Every time, all the synchronous generators are shown as a model of a classical generator and constant impedance models for each of the loads. MATLAB® 2022b is used to execute all simulations on a personal computer equipped with an Intel Core i5 3.1 GHz processor and 8GB of RAM. With an integration time step of 0.01s, the total simulation duration is fixed at 3.0s. The terminating condition is set at 100 iterations.

Test system I (WSCC 3-machine 9-bus system)

Case I: Base loading conditions (OPF without transient stability constraint) The infrastructure consists of 11 transmission lines, 2 regulation transformers, 2 shunt converters, 9 buses, and 3 producing units, as illustrated in Figure 11.1 (Nguyen and Pai [39]): this is where the information for the system bus, the lines, the generator capacity, and the cost coefficients is taken from. On the other hand, the voltage and current that are viewed as generator parameters (aside from slack bus-1) are regarded as continuous variables, and the VAR and regulation transformers react to the prompt injections of the shunt capacitors. The range of [0.95 p.u., 1.1 p.u.] and [0.95 p.u., 1.05 p.u.] (Nguyen and Pai [39]) for the respective voltage

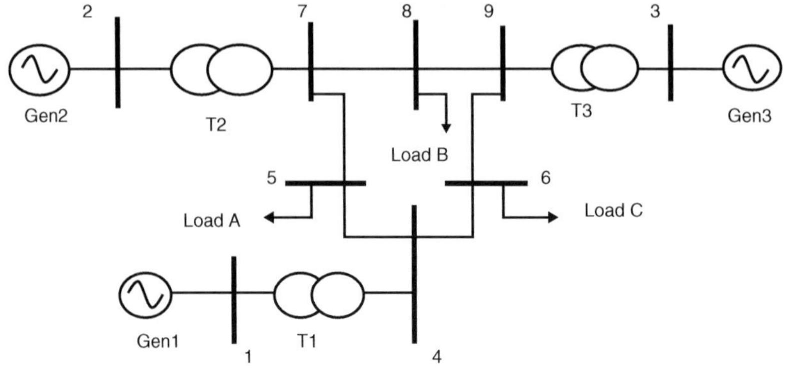

Figure 11.1 The WSCC 3-machine 9-bus test system.

amplitudes of the generators and load buses is considered. The range of available regulating transformer tap options is limited to [0.9 p.u. and 1.1 p.u.]. Table 11.1 displays the simulation results for the fuel price fitness function for the VAR injection constraints of the shunt capacitors, which are in the range of [0 Mvar, 5 Mvar], with a unique initial value of 0.5 Mvar. Comparing the computed findings to modified differential evolution (MDE) [40], BBO [41], GSA [42], and teaching learning-based optimization (TLBO) [43] has been done. Compared to the additional techniques in Table 11.1, the minimal objective function values for CQODSA and DSA are 1131.0112 and 1131.0067, respectively. Therefore, Table 11.1 suggests that the CQODSA and DSA algorithms offer better controller parameters for Case I when compared to other methods.

Case II: OPF with transient stability constraint (**3 phase to ground fault at bus** 7) Specifically, an asymmetric, three-phase to ground is presented in between lines 7–5, and at bus number 7, which is shown in Figure 11.2. This figure represented that the suggested algorithms are appropriate for managing the power systems. This was resolved by briefly opening the circuit

Table 11.1 Simulation outcomes achieved by different methodologies for Case I (test system I)

Algorithms	*ABC*	*DE*	*TS*	*TDS*	*TLBO*	*QODSA*	*CQODSA*
PG_1 (MW)	117.69	130.94	170.20	117.85	105.310	104.67	107.56
PG_2 (MW)	105.89	94.46	48.94	103.50	113.33	112.42	111.26
PG_3 (MW)	94.23	93.09	98.74	96.66	99.24	97.87	98.42
VG_1 p.u	1.025	0.9590	1.000	1.05	1.0988	1.0962	1.0865
VG_2 p.u	1.070	1.0133	1.000	1.05	1.0987	1.0945	1.0966
VG_3 p.u	1.070	1.0467	1.000	1.04	1.099	1.0932	1.0852
FC ($/hour)	1133.18	1140.06	1179.95	1134.01	1131.19	1128.47	1127.4227

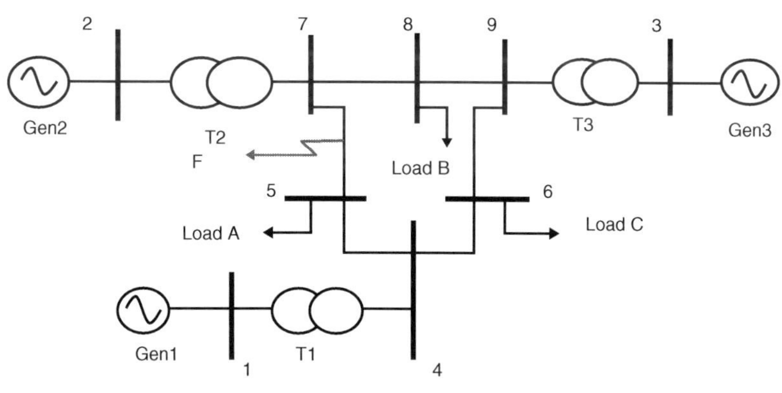

Figure 11.2 The 3-machine, 9-bus WSCC test system has a problem at bus 7.

Table 11.2 Simulations outcomes achieved by various methodologies for Case II (WSCC 9-bus system)

Algorithms	*ABC*	*DE*	*TS*	*TDS*	*TLBO*	*QODSA*	*CQODSA*
PG_1 (MW)	105.94	107.15	105.94	106.19	105.94	107.89	107.65
PG_2 (MW)	113.04	114.18	113.04	112.96	113.04	113.99	112.78
PG_3 (MW)	99.23	96.74	99.29	99.20	99.24	98.89	99.67
VG_1 p.u	1.05	1.033	1.050	1.000	1.05	1.065	1.076
VG_2 p.u	1.05	1.024	1.050	1.000	1.05	1.04	1.087
VG_3 p.u	1.04	1.028	1.040	1.000	1.04	1.07	1.06
FC ($/hour)	1132.17	1131.87	1132.30	1132.59	1132.18	1131.67	1126.9643

breaker contact (0.35 s). The result in this example satisfies the transient stability limit. ABC [44], dynamic variation (DV) [35], sensitive to trend line (TS) [35], simulations in spatial domain [35], and TLBO [43] are the least full cost values that are compared in Table 11.2. The result shows how effective the suggested CQODSA is at identifying the best overall solution. The suggested CQODSA technique yields a dependable and secure solution.

Test system II (New England 10-machine 39-bus system)

The recommended methods are also carried out using the New England 10-machine 39-bus test equipment, as shown in Figure 11.3. Within the network, there are 19 load buses and 10 generator buses. Bus 39 will take on the role of the load bus. The voltage magnitude spans of the generators and the load buses are [0.95 p.u., 1.1 p.u.] and [0.95 p.u., 1.05 p.u.], respectively. There are restrictions placed on [0.9 p.u. and 1.1 p.u.], [0.9 p.u., 1.1 p.u.], [0.9 p.u., 1.1 p.u.], [0.9 p.u., 1.1 p.u.], and [0.9 p.u., 1.1 p.u]. When utilizing shunt capacitors, the VAR insertion limits are considered for the [0 Mvar, 5 Mvar] span with a discrete value of 0.5 Mvar movements. Reducing the overall cost of fuel is the main goal. Active production and voltage are shown in simulation results 3–4 for the wellness feature (fuel price).

A three-phase asymmetrical fault at bus 29 and across lines 28 and 29 is present in the system. The above problem can be fixed by momentarily opening the connections on the circuit breakers. The classical model (CM) [45], the dynamic model (DM) [45], the variables simulation method (DSA) [46], prototype, and TLBO [43] were used for calculating the actual power generation and voltage of various generating units, as well as the fuel price and transmission loss. The wellness feature values are contrasted with the strategies outlined in the literature. When compared to the other algorithms in Table 11.4, CQODSA yields the lowest minimum objective function value, 41.1290, according to an examination of the table. Table 11.4

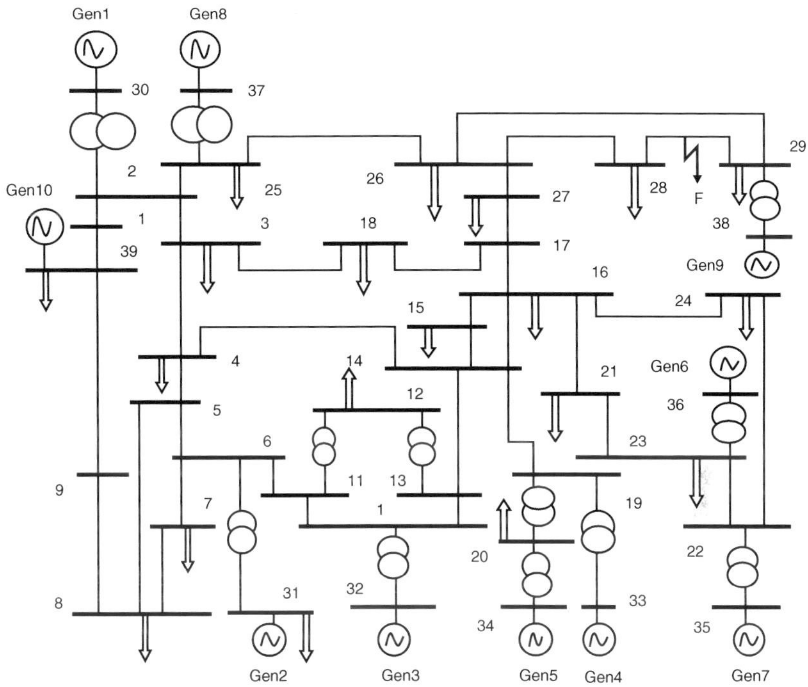

Figure 11.3 The 10-machine, 39-bus New England test system.

thus suggests that the CQODSA and quasi oppositional differential search algorithm (QODSA) algorithms, as displayed in Tables 11.3 and 11.4, provide controller settings that are more suitable than those of other methods. Table 11.4 provides data on fuel consumption, energy losses as assessed using various techniques, voltage, and actual power generation of various producing units. This leads us to the conclusion that the fuel cost obtained by using the CQODSA and QODSA approaches that are offered is substantially lower than the cost obtained by applying other methods.

CONCLUSION

This chapter has introduced, developed, and effectively used several TSC-OPF problem-solving methodologies. Both the New England 10-generator 39-bus system and the WSCC 3-generator 9-bus system are realistic generators running on two test devices assessed using the methodology that is given. Based on the simulation results generated by different algorithms, it can be concluded that the proposed CQODSA and QODSA algorithms find the optimal solutions in all scenarios while staying within operating restrictions.

Table 11.3 Simulation outcomes achieved by different methodologies (test system II)

Active power generation (MW)	*TLBO*	*CM*	*DM*	*DSA*	*QODSA*	*CQODSA*
PG30	244.22	248.7300	249.4470	247.8300	245.8760	**245.6578**
PG31	560.07	577.8400	578.3590	577.2300	587.2133	**567.3214**
PG32	643.63	654.4700	654.3560	653.4100	667.9804	**654.1265**
PG33	632.98	645.0000	641.7600	643.2800	636.2311	**645.9830**
PG34	510.98	518.8200	517.4100	517.7800	520.4318	**541.9807**
PG35	653.06	664.3200	660.7310	662.4600	676.5462	**657.3248**
PG36	561.03	571.3700	568.1810	569.5900	545.9810	**547.9801**
PG37	532.86	547.8100	547.6950	543.8800	545.3450	**567.1231**
PG38	840.15	752.0200	754.6180	774.5400	758.1234	**768.1326**
PG39	958.45	995.6000	1003.1810	1000.3500	998.1231	**980.1358**
Cost ($/hour)	60873.5574	61600.7600	61597.7600	61799.6800	61689.4288	**61434.9800**

Bold indicates that best output.

Table 11.4 Simulation outcomes achieved by different methodologies (test system II)

Generated voltages (p.u.)	*TLBO*	*CM*	*DM*	*DSA*	*QODSA*	*CQODSA*
V30	1.0996	1.0150	1.0150	0.9840	0.9865	0.9898
V31	1.0924	1.0870	1.0870	1.0740	1.0651	1.0789
V32	1.0863	1.0290	1.0290	1.0080	1.0085	1.0901
V33	1.0985	1.0160	1.0160	1.0140	1.4310	1.6511
V34	1.0997	1.0220	1.0220	1.0190	1.0187	1.0956
V35	1.0999	1.0620	1.0620	1.0670	1.0679	1.0789
V36	1.0999	1.0900	1.0900	1.0870	1.0967	1.0878
V37	1.0994	1.0470	1.0470	1.0120	1.0165	1.0123
V38	1.0995	1.0380	1.0380	1.0510	1.0678	1.0234
V39	1.0998	1.0530	1.0530	1.0190	1.0912	1.0966
Loss (MW)	43.15359	NA	NA	NA	44.1324	43.1351

To substantiate the superiority, it can be said that QODSA and CQODSA greatly outperformed all other classifiers discussed in this chapter in terms of confirming the efficiency, rating, and continuity of the final technique. The methods also exhibit strong convergence qualities. Thus, for the purpose of overall optimization of any TSC-OPF challenges based on originality, the suggested CQODSA and QODSA algorithms are very interesting evolutionary optimization approaches. The suggested algorithms can still be improved upon to yield the most likely optimal response in a reasonable amount of time, even if they generate better results.

REFERENCES

1. HW Dommel and WF Tinney. Optimal power flow solutions. *IEEE Transactions on Power Apparatus and Systems*, 10:1866–1876, 1968.
2. E Lobato, L Rouco, MI Navarrete, R Casanova, and G Lopez. An lp-based optimal power flow for transmission losses and generator reactive margins minimization. In: *2001 IEEE Porto Power Tech Proceedings (Cat. No. 01EX502)*, vol. 3, p. 5. IEEE, 2001.
3. JA Momoh. A generalized quadratic-based model for optimal power flow. In: *Conference Proceedings. IEEE International Conference on Systems, Man and Cybernetics*, pp. 261–271. IEEE, 1989.
4. JA Momoh and JZ Zhu. Improved interior point method for opf problems. *IEEE Transactions on Power Systems*, 14(3):1114–1120, 1999.
5. DI Sun, B Ashley, B Brewer, A Hughes, and WF Tinney. Optimal power flow by newton approach. *IEEE Transactions on Power Apparatus and Systems*, 10:2864–2880, 1984.

6. X Yan and VH Quintana. Improving an interior-point-based opf by dynamic adjustments of step sizes and tolerances. *IEEE Transactions on Power Systems*, 14(2):709–717, 1999.
7. JA Momoh, ME El-Hawary, and R Adapa. A review of selected optimal power flow literature to 1993. ii. newton, linear programming and interior point methods. *IEEE Transactions on Power Systems*, 14(1):105–111, 1999.
8. D Pudjianto, S Ahmed, and G Strbac. Allocation of var support using lp and nlp based optimal power flows. *IEE Proceedings-Generation, Transmission and Distribution*, 149(4):377–383, 2002.
9. PK Roy and D Mandal. Quasi-oppositional biogeography-based optimization for multi-objective optimal power flow. *Electric Power Components and Systems*, 40(2):236–256, 2011.
10. AA Abou El-Ela and MA Abido. Optimal operation strategy for reactive power control. *Modelling Simulation and Control a General Physics Matter and Waves Electrical and Electronics Engineering*, 41:19–19, 1992.
11. A Jr Santos and GRM Da Costa. Optimal-power-flow solution by Newton's method applied to an augmented lagrangian function. *IEE Proceedings Generation, Transmission and Distribution*, 142(1):33–36, 1995.
12. EP De Carvalho, A dos Santos Júnior, and TF Ma. Reduced gradient method combined with augmented lagrangian and barrier for the optimal power flow problem. *Applied Mathematics and Computation*, 200(2):529–536, 2008.
13. F Capitanescu and L Wehenkel. Experiments with the interior-point method for solving large scale optimal power flow problems. *Electric Power Systems Research*, 95:276–283, 2013.
14. Z-L Gaing and R-F Chang. Security-constrained optimal power flow by mixed-integer genetic algorithm with arithmetic operators. In *2006 IEEE Power Engineering Society General Meeting*, p. 8. IEEE, 2006.
15. LL Lai, JT Ma, R Yokoyama, and M Zhao. Improved genetic algorithms for optimal power flow under both normal and contingent operation states. *International Journal of Electrical Power & Energy Systems*, 19(5):287–292, 1997.
16. CA Roa-Sepulveda and BJ Pavez-Lazo. A solution to the optimal power flow using simulated annealing. *International Journal of Electrical Power & Energy Systems*, 25(1):47–57, 2003.
17. MA Abido. Optimal power flow using tabu search algorithm. *Electric Power Components and Systems*, 30(5):469–483, 2002.
18. AK Swain and AS Morris. A novel hybrid evolutionary programming method for function optimization. In: *Proceedings of the 2000 Congress on Evolutionary Computation. CEC00 (Cat. No.* 00TH8512*)*, vol. 1, pp. 699–705. IEEE, 2000.
19. J Cai, X Ma, L Li, Y Yang, H Peng, and X Wang. Chaotic ant swarm optimization to economic dispatch. *Electric Power Systems Research*, 77(10):1373–1380, 2007.
20. M Tripathy and S Mishra. Bacteria foraging-based solution to optimize both real power loss and voltage stability limit. *IEEE Transactions on Power Systems*, 22(1):240–248, 2007.

21. AA Abou El Ela, MA Abido, and SR Spea. Optimal power flow using differential evolution algorithm. *Electric Power Systems Research*, 80(7):878–885, 2010.
22. S Sivasubramani and KS Swarup. Multiagent based differential evolution approach to optimal power flow. *Applied Soft Computing*, 12(2):735–740, 2012.
23. PK Roy, SP Ghoshal, and SS Thakur. Biogeography based optimization for multi-constraint optimal power flow with emission and non-smooth cost function. *Expert Systems with Applications*, 37(12):8221–8228, 2010.
24. AH Khazali and M Kalantar. Optimal reactive power dispatch based on harmony search algorithm. *International Journal of Electrical Power & Energy Systems*, 33(3):684–692, 2011.
25. K Ayan and U Kılıç. Artificial bee colony algorithm solution for optimal reactive power flow. *Applied Soft Computing*, 12(5):1477–1482, 2012.
26. PK Roy, B Mandal, and K Bhattacharya. Gravitational search algorithm based optimal reactive power dispatch for voltage stability enhancement. *Electric Power Components and Systems*, 40(9):956–976, 2012.
27. M Basu. Multi-objective optimal power flow with facts devices. *Energy Conversion and Management*, 52(2):903–910, 2011.
28. H Nobahari, M Nikusokhan, and P Siarry. A multi-objective gravitational search algorithm based on non-dominated sorting. *International Journal of Swarm Intelligence Research (IJSIR)*, 3(3):32–49, 2012.
29. MA Abido and JM Bakhashwain. Optimal var dispatch using a multiobjective evolutionary algorithm. *International Journal of Electrical Power & Energy Systems*, 27(1):13–20, 2005.
30. D Gan, RJ Thomas, and RD Zimmerman. Stability-constrained optimal power flow. *IEEE Transactions on Power Systems*, 15(2):535–540, 2000.
31. Y Yuan, J Kubokawa, and H Sasaki. A solution of optimal power flow with multicontingency transient stability constraints. *IEEE Transactions on Power Systems*, 18(3):1094–1102, 2003.
32. L Chen, YT, H Okamoto, R Tanabe, and A Ono. Optimal operation solutions of power systems with transient stability constraints. *IEEE Transactions on Circuits and Systems I: Fundamental Theory and Applications*, 48(3):327–339, 2001.
33. D Karaboga et al. An idea based on honey bee swarm for numerical optimization. Technical Report, Technical Report-tr06, Erciyes University, Engineering & Computer Faculty, 2005.
34. KY Chan, KW Chan, SH Ling, HHC Iu, and GTY Pong. Solving multicontingency transient stability constrained optimal power flow problems with an improved ga. In: 2007 *IEEE Congress on Evolutionary Computation*, pp. 2901–2908. IEEE, 2007.
35. HR Cai, CY Chung, and KP Wong. Application of differential evolution algorithm for transient stability constrained optimal power flow. *IEEE Transactions on Power Systems*, 23(2):719–728, 2008.
36. T Athay, R Podmore, and S Virmani. A practical method for the direct analysis of transient stability. *IEEE Transactions on Power Apparatus and Systems*, (2):573–584, 1979.

37. P Civicioglu. Transforming geocentric cartesian coordinates to geodetic coordinates by using differential search algorithm. *Computers & Geosciences*, 46:229–247, 2012.
38. HR Tizhoosh. Opposition-based learning: a new scheme for machine intelligence. In: *International Conference on Computational Intelligence for Modelling, Control and Automation and International Conference on Intelligent Agents, Web Technologies and Internet Commerce (CIMCA-IAWTIC'06)*, vol. 1, pp. 695–701. IEEE, 2005.
39. TB Nguyen and MA Pai. Dynamic security-constrained rescheduling of power systems using trajectory sensitivities. *IEEE Transactions on Power Systems*, 18(2):848–854, 2003.
40. NJ Vickers. Animal communication: when i'm calling you, will you answer too? *Current Biology*, 27(14):R713–R715, 2017.
41. A Bhattacharya and PK Chattopadhyay. Application of biogeography-based optimisation to solve different optimal power flow problems. *IET Generation, Transmission & Distribution*, 5(1):70–80, 2011.
42. S Duman, U Güvenç, Y Sönmez, and N Yörükeren. Optimal power flow using gravitational search algorithm. *Energy Conversion and Management*, 59:86–95, 2012.
43. A Mukherjee, S Paul, and PK Roy. Transient stability constrained optimal power flow using teaching learning-based optimization. *International Journal of Energy Optimization and Engineering (IJEOE)*, 3(4):55–71, 2014.
44. K Ayan and U Kiliç. Solution of transient stability-constrained optimal power flow using artificial bee colony algorithm. *Turkish Journal of Electrical Engineering and Computer Sciences*, 21(2):360–372, 2013.
45. H Ahmadi, H Ghasemi, AM Haddadi, and H Lesani. Two approaches to transient stability-constrained optimal power flow. *International Journal of Electrical Power & Energy Systems*, 47:181–192, 2013.
46. X Tu, L-A Dessaint, and H Nguyen-Duc. Transient stability constrained optimal power flow using independent dynamic simulation. *IET Generation, Transmission & Distribution*, 7(3):244–253, 2013

Chapter 12

Illuminating the path to a sustainable future by harnessing AI-powered renewable energy systems

Nipun Jain, Yusuf Olatunji Waidi, Samir Das, Ranjit Barua, Vilay Vannaladsaysy, and Sudipto Datta

INTRODUCTION

Integrating complex artificial intelligence (AI) into smart energy systems presents a complicated challenge, and it demands technical and social expertise. This intricate subject requires a clearly defined scope and a well-formulated research problem for better development. As the pursuit of sustainable development unites society, academia, and industry in finding solutions for a post-industrial future, the associated challenges and their evolving literature continue to enrich our understanding of this crucial transition.

The global energy landscape is undergoing a significant transformation, driven by the pressing need for increased flexibility, lower costs, and reduced environmental impact. Renewable energy (RE) has emerged as a key player in this shift, offering a sustainable and resilient future in the face of climate change and resource depletion [1]. RE's journey began during the energy crisis of the 1970s, stimulated by concerns over diminishing conventional fuel reserves. This sparked efforts to develop renewable alternatives and conserve natural resources. In the following decades, growing awareness of pollution, global warming, and resource depletion further intensified the call for clean energy solutions [2]. Today, with our focus on reducing carbon emissions and protecting public health, the role of RE is more crucial than ever. Integrating RE sources is crucial to tackling dual challenges, such as reducing greenhouse gas emissions. It can offer inherent sustainability due to its natural regeneration, making it a tangible solution for a cleaner future [3]. However, fluctuating energy output from renewables poses a significant challenge in managing supply to meet the energy needs cleanly and reasonably. The government of a country will have to face economic and environmental concerns that demand innovative solutions. Fortunately, AI has risen as an investigative platform to address these needs. Its ability to perform high-speed predictions and adaptations in changing conditions provides insightful explanations, making it ideal for optimizing RE management [4].

DOI: 10.1201/9781003628279-12

The emergence of AI demands novel approaches for organizing activities and infrastructure within the RE sector. Effectively navigating the challenges will boost its growth. The resilience within the system will require advancements in energy infrastructure design, RE deployment, and production methods. Our analysis delves into several key areas to understand AI's impact on the RE sector. We analyze the efficiency of RE transformation throughout the energy chain, exploring how AI can optimize processes from source to final consumption. We investigate the potential shift in the distribution of RE sources like solar, wind, and biomass due to AI advancements. Additionally, we compare labor productivity in the RE sector to the broader economy, assessing its correlation with investment levels and exploring how AI might influence this dynamic. Finally, we consider the potential of AI integration within future smart city research, evaluating its contribution to the RE sector's overall efficiency, adaptability, and role in smart city development.

INTEGRATION OF RE AND AI FOR ENERGY MANAGEMENT

The global energy sector is undergoing a major transformation driven by a rise in RE technologies. This shift is driven by the urgency to address climate change, diminishing fossil fuel reserves, and rising pollution. With its vast potential and independence from volatile fossil fuel prices, RE offers a powerful solution for future development [5]. However, integrating these variable sources presents unique challenges. First, RE's intermittent nature necessitates advanced technologies to handle fluctuating power generation, large data volumes, and bidirectional energy flows. This requires grid upgrades to ensure high power quality and efficient transmission, especially in remote areas where traditional fossil fuel-based generation is less feasible [6]. Second, the variability demands innovative solutions for energy storage and accurate forecasting to optimize utilization and avoid reliance on backup fossil fuels [7] (Figure 12.1).

AI systems can change their behavior without explicitly having to be reprogrammed; instead, they rely simply on the experience they have accumulated from seeing, collecting, and evaluating enormous volumes of data [8]. They offer a summary of steps, and the algorithms can extract valuable information. AI is required to support decision-making as the power systems are becoming more complicated because of the volume and complexity of data they hold. It has the potential to alter the world's energy network [9]. The unpredictability of solar or wind radiation is a cause of concern for grid operators because RE sources make up a bigger share of their energy structures. Therefore, they need weather estimates, including

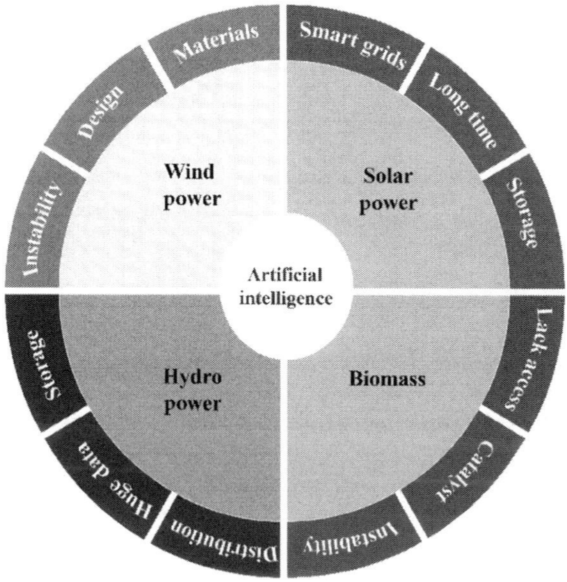

Figure 12.1 Integrating artificial intelligence and renewable energy sources will help extract maximum energy and consumption.

solar and wind, to determine how much energy they will contribute to the grid over the next period [10]. Since it's essential for both the optimal control of energy flows into RE sources and the smooth operation of the power grid, probabilistic load forecasting has gained importance in energy system planning and management [11]. Forecasts of load and RE are necessary for effective energy management. Variable energy sources and shifting loading capacities, which were not intended for the real energy grid, can be accommodated by RE sources [12]. It is evolving because of the continuous addition of new, advanced algorithms for generating, managing, transmitting, and utilizing RE. The new methods have improved accuracy and need less computation time. To overcome an algorithm's constraints, employing the hybridization of two or more algorithms becomes essential [13]. The combination of AI and RE will revolutionize the energy sector and boost domestic and global sustainability. AI has the potential to improve the efficiency of the real estate industry via pattern recognition and prediction, autonomous job execution without human guidance, supply optimization, and improved decision-making [14]. This offers improved visions into procedures because of quicker predicting and more intelligent connections between crucial mechanisms due to the fast growth of AI-infused technology.

RECENT ADVANCEMENTS IN AI USAGE

The aim to mitigate the effects of climate change is fueling a worldwide surge in the consumption of RE sources. RE is a viable answer to the modern economy and an important field for research and development, given the continuous depletion of natural resources and the need to protect the environment [15]. AI is increasingly being incorporated into this industry, improving RE outcomes in terms of accessibility and efficiency. It can facilitate the management of consumption and energy production in a dynamic market and environment [16]. The unpredictable nature of these energy sources is a major problem for this industry and is getting worse. AI presents a viable solution to mitigate the variability risk of RE sources. This can be achieved through applications that reduce storage costs, perform predictive analysis, identify patterns, and improve user-grid connectivity, ultimately contributing to grid sustainability, reliability, and stability [17] (Figure 12.2).

AI will usher in a new age of social awareness and information-driven global growth. It suggests new methods for using conventional production elements and new guidelines for allocating tasks to address present problems. While the historical history of AI must demonstrate notable turning points, current advances in machine learning (ML), cognitive computing, and neural network research are opening up previously unimagined possibilities for monitoring, controlling, and optimizing energy [18]. Future generation smart city services might benefit greatly from combining AI with sensor networks. A sub-section of computer science will emphasize generating the future technical advances that will give human-like capabilities to computers. Because there is an oversupply of computing power and the notable rise in data quantities, it has been employed more frequently. The Internet will drive the expansion of AI as the data can be

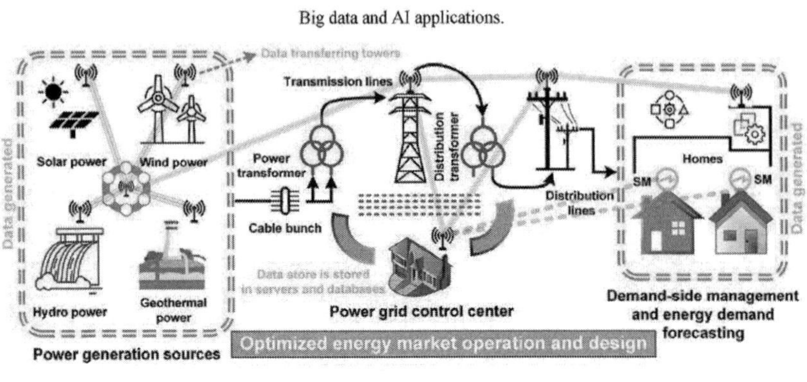

Figure 12.2 AI-based big data applications with renewable energy sources. (Reproduced from Ref. [16].)

handled more easily, enabling fast and effective decision-making. Its ability to support system standardization will customize and parametrize production and consumption in the energy industry [19]. In the energy industry, envisioning smart agents driven by AI algorithms and smart grid technology necessitates social and economic research that goes beyond technological constraints. An eco-friendly AI in the energy division that supports an AI-enabled smart energy grid may be supported by various AI techniques, including ML, data mining, and structured data management technology. It must be included at every level to continue the evolution of energy substructures. Enhancing the architecture and the use and generation of RE are imperative [20]. AI has become crucial due to the energy industry's reliance on big data and ever-more sophisticated systems. It has the potential to significantly boost the clean power industry, particularly via improved tracking, operations, restoration, and storage for green energies, as well as efficient operational efficiency and management. These primary uses include the integration of clean power with power grids for (1) clean power generations, taking into account supplies instability with the unpredictability of green energies; (2) safer processes with energy stability and efficiency; (3) precise consumption and climate forecasting; (4) effective control of the market sides; (5) functions related to power storing; (6) market structure with functioning; (7) improved communication among micro-grids and system parts [21].

The innovative disruption hypothesis, which emphasizes their significance with technology-driven development, explains today's market leadership in the real estate industry that higher levels of AI mark rising technologies. The implicit relationships between AI and its potential uses are shown in Figure 12.2. For a managerial concept, innovative, disruptive technology describes the choices made by dominant businesses in response to various forms of technology and economic disruption. Although innovative disruption is originally based on recognized or customary measurements, it can introduce an innovative set of unique characteristics. Technological disruptions have the potential to advance, dislodging rivals or emerging as opponents, although their initial impact is thought to be limited to easy applications at the bottom of existing industries [22]. Manufacturers' adaptability in implementing disruptive technology is contingent upon the management strategy implemented on novel technological advances: possibility or danger. When disruptive technology is viewed as a danger, more resources are allocated. Manufacturers who address the requirements for low-demanding consumer groups or smaller market groups are the ones who demonstrate disruption [23]. The disruption happens once new competitors start drawing in mainstream clients as an outcome of established manufacturers concentrating on their most valuable market niches (Figure 12.3).

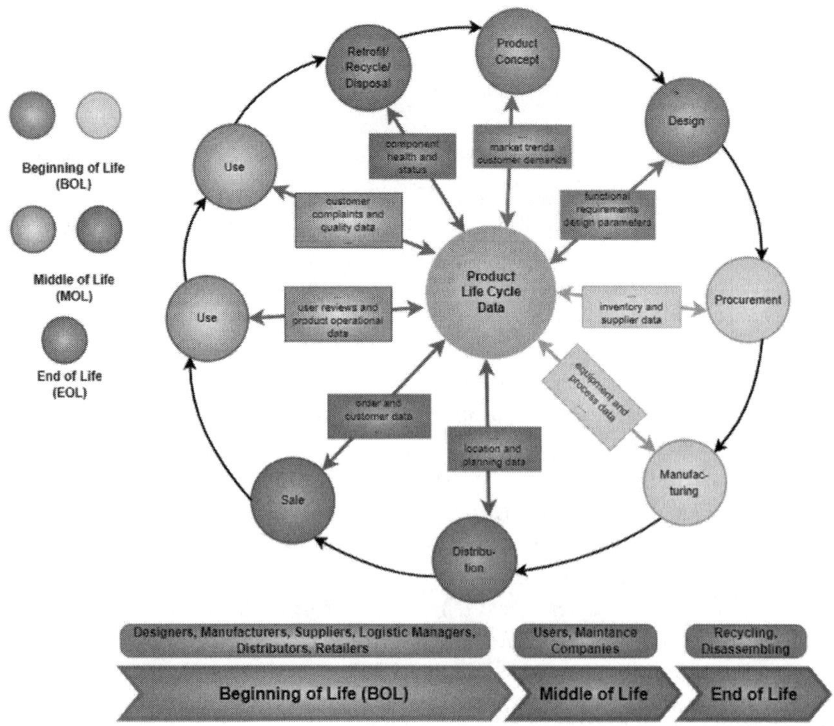

Figure 12.3 Over a product's lifetime, a significant amount of diversified and high-dimensional data is produced that needs to be managed by AI for better advancement. (Reproduced from Ref. [19].)

AI-POWERED RENEWABLE ENERGY SOLUTIONS WITH SIGNIFICANT SOCIAL EFFECTS AND COST-EFFECTIVENESS

The main goal of this field was to investigate this multifaceted link. We aim to make a valuable contribution to the scientific discourse regarding how AI can improve the effectiveness of the clean energy industry, with a focus on sustainability and financial effectiveness, by utilizing an innovative technique. Within the framework for this study, we add a new and noteworthy topic to the study goal for the efficacy of energy improvements. The accessibility of sources and workers' output needs to be acknowledged to start a new study debate in such a complex and multifaceted study framework [24]. Because it depends on the unpredictable ecosystem, generating energy from natural resources is unpredictable. The growing use of renewable energies in the power industry highlights the need for precise forecasting of variable resources. In order to distribute required amounts in a manner that maintains grids' balance and minimizes the impact of variations on stability, predicted errors must be minimized. Outages in supplies may result in additional manufacturer expenses [25]. In reaction to variations in energy supplies, demands are incredibly rigid, and users need to modify demands in accordance with their requirements and actions rather than the state of the market or the surrounding circumstances [26]. Correspondingly, the solutions for balancing the variable and uncertain supply of RE are, on the one hand, other power generators and improved control regarding supplies (although lacking in respect to costs, accessibility, or environmental pressure) such as gas and coal while simultaneously incorporating AI [27]. It is important to consider the price of utilizing any of those options with the overall (possibilities) value of utilizing energy. According to specialists, AI can provide judgments, suggestions, and predictions about a particular set of goals defined by humans, impacting the actual or virtual world [28]. Due to their ability to predict outcomes based on historical data analysis, oversee and manage operations, make prompt decisions, and store energy, AI-powered technologies in the real estate industry can potentially increase investment return rates while concurrently lowering costs and improving usability [29]. Advantages entail expenses and hazards related to system security and adverse possessions on the labor strength. The application of AI will contribute to economic development while considering the implications on the environment and human well-being. This will enable economic growth to exceed the limitations of conventional production forces. This strategy is especially obvious when applying AI to the real estate industry [30]. AI will alter the technologies employed and the structure of the demand for work, necessitating investments in lifelong learning and technological capital to reshape the labor market and industrial sector in light of the new nature of processes [31]. From this vantage point, the labor market presents the real estate industry with fresh difficulties in skills and labor shortages. Developed nations have an advantage in investments and the labor market since they have a large pool of highly skilled individuals. AI has substantial potential for creating new job opportunities despite the changes in occupational structure, which is a major policy problem in many nations [32]. Although AI suggests more complicated and faster decision-making than a human operator, humans are still at the core of the processes because they can oversee, program, reconfigure, and enforce rules and safety operating guidelines, among other things. A further significant component of this system is its interpretive method of crucial information-gathering within the energy industries. This investigation also provides a qualitative strategy toward deploying AI as an innovative technology that supports and allows the establishment of a fresh field of smart energy systems as a facility in RE.

THE ROLE OF AI IN THE DEVELOPMENT OF SMART CITIES

Modern scientific inheritance, shown by the success of transformation in procedures and modifications in the developmental framework, is the precursor to the need for AI. The field of RE is gaining more and more attention. The examination of economic data might yield important findings about the need for continued financial support and incentives for the RE sector [33]. Additionally, given the RE sector's close links to sustainability and economic development, it is necessary to analyze the variables that support the effectiveness and effect of society, particularly from the perspective of modern city exploration. In the context of smart energy grids, the AI domain appears to be a major enabler of unanticipated efficiency possibilities, thanks to its sophisticated ML and intelligent computing. In the RE sector, conventional energy technologies continue to rule. Emergent technologies that use AI are predicted to develop in efficiency and performance over time, diffuse more quickly, and eventually take over the RE market, as seen by numerous industries [34]. The sequence in which technology enters the market determines its success in the rivalry between established and new technologies in the real estate sector. As latecomers to the RE industry, emerging technologies have this disadvantage. The consistency with consumers' tastes, the level of pricing, and their performance ratings will counterbalance it.

The combination of cutting-edge information systems research with social science research is the primary contribution. Future smart city design will face problems, from examining the effects of RE adoption on the economy to enhancing it using advanced AI technology [35]. The many components of this research study support a framework for examining how AI development affects RE smart energy systems. AI can recognize cyclical and repeating models and samples, recognize discrepancies in process stages, predict trends for both energy fabrication and requirement, alleviate or eliminate imbalances between contribution and demand caused by fluctuations in RE sources, enhance effectiveness in energy usage, and contribute to the cost-effective optimization of power grids [36]. It will help make decisions through advanced analysis of extensive datasets, utilizing forecast, model, and administration techniques. Furthermore, it can autonomously make decisions in comparable or unfamiliar scenarios, drawing from earlier experiences or incorporating past conclusions into the accessible information.

Approaching the Upcoming Advanced Cities Study to incorporate RE with AI effectiveness will involve important factors in the progression from manufacture to use. Incorporation at both the macro and micro scales encompasses solicitations for intelligent homes and incorporated frameworks for maximizing energy smart network usage. Integrating ecological considerations into global service design aims to offer dynamic and comprehensive solutions. AI will additionally allow global energy systems to be immediately utilized in addition to developing sophisticated statistics.

CONCLUSION

The use of AI represents one of the many difficult areas of future worldwide study, especially within the RE arena. AI and ML have allowed solutions to crucial issues like efficiency, the production of smart adjustable systems, and complicated mathematical techniques and methodologies. AI–ML has a large history of usage in diagnostic, prognostic, and predictive aspects for enhancing clinical work and patient care. ML has been used to construct high-dimensional features. Advances in this field have produced a recognizable set of unique systems that employ prediction models to gather copious amounts of data in order to guide the subsequent wave of research endeavors [37]. Technical creativity in AI harnesses the renewable segment by providing efficiency and rectifying power usage. Manufacturers in the green energy sector profit from technological advancements, which are also important for the nation to reduce dependence on traditional energy sources. AI tools provide a disruptive innovation source in the energy market that utilizes various methods for ML and computer programs. These allow decisions for economic and social benefits employed in various areas. Systems with AI are used to encourage ecological accountability for the environment. It symbolizes the next phase in the RE industry, either literally or comparatively, given that the development and presentation of effective procedures work seamlessly. We therefore conclude that AI has the potential to become prevalent in conservative energy techniques.

REFERENCES

1. Burger, J.; Gochfeld, M. A Conceptual Framework Evaluating Ecological Footprints and Monitoring Renewable Energy: Wind, Solar, Hydro, and Geothermal. *Energy & Power Engineering* **2012**, *4*(4), 303–314. https://doi.org/10.4236/epe.2012.44040.
2. Turner, J. A. A Realizable Renewable Energy Future. *Science* (1979) **1999**, *285*(5428), 687–689. https://doi.org/10.1126/science.285.5428.687.
3. Puri, V.; Jha, S.; Kumar, R.; Priyadarshini, I.; Hoang Son, L.; Abdel-Basset, M.; Elhoseny, M.; Viet Long, H. A Hybrid Artificial Intelligence and Internet of Things Model for Generation of Renewable Resource of Energy. *IEEE Access* **2019**, *7*, 111181–111191. https://doi.org/10.1109/ACCESS.2019.2934228.
4. Mellit, A.; Kalogirou, S. A. Artificial Intelligence Techniques for Photovoltaic Applications: A Review. *Progress in Energy and Combustion Science* **2008**, *34*(5), 574–632. https://doi.org/10.1016/j.pecs.2008.01.001.

5. Khokhar, S.; Mohd Zin, A. A. B.; Mokhtar, A. S. B.; Pesaran, M. A Comprehensive Overview on Signal Processing and Artificial Intelligence Techniques Applications in Classification of Power Quality Disturbances. *Renewable and Sustainable Energy Reviews* **2015**, *51*, 1650–1663. https://doi.org/10.1016/j.rser.2015.07.068.
6. Gupta, A.; Chauhan, Y. K.; Pachauri, R. K. A Comparative Investigation of Maximum Power Point Tracking Methods for Solar PV System. *Solar Energy* **2016**, *136*, 236–253. https://doi.org/10.1016/j.solener.2016.07.001.
7. Sharma, V.; Cortes, A.; Cali, U. Use of Forecasting in Energy Storage Applications: A Review. *IEEE Access* **2021**, *9*, 114690–114704. https://doi.org/10.1109/ACCESS.2021.3103844.
8. Azad, H. B.; Mekhilef, S.; Ganapathy, V. G. Long-Term Wind Speed Forecasting and General Pattern Recognition Using Neural Networks. *IEEE Transactions on Sustainable Energy* **2014**, *5*(2), 546–553. https://doi.org/10.1109/TSTE.2014.2300150.
9. Das, U. K.; Tey, K. S.; Seyedmahmoudian, M.; Mekhilef, S.; Idris, M. Y. I.; Van Deventer, W.; Horan, B.; Stojcevski, A. Forecasting of Photovoltaic Power Generation and Model Optimization: A Review. *Renewable and Sustainable Energy Reviews* **2018**, *81*, 912–928. https://doi.org/10.1016/j.rser.2017.08.017.
10. Hong, T.; Fan, S. Probabilistic Electric Load Forecasting: A Tutorial Review. *International Journal of Forecasting* **2016**, *32*(3), 914–938. https://doi.org/10.1016/j.ijforecast.2015.11.011.
11. Voyant, C.; Notton, G.; Kalogirou, S.; Nivet, M.-L.; Paoli, C.; Motte, F.; Fouilloy, A. Machine Learning Methods for Solar Radiation Forecasting: A Review. *Renew Energy* **2017**, *105*, 569–582. https://doi.org/10.1016/j.renene.2016.12.095.
12. Diamantoulakis, P. D.; Kapinas, V. M.; Karagiannidis, G. K. Big Data Analytics for Dynamic Energy Management in Smart Grids. *Big Data Research* **2015**, *2*(3), 94–101. https://doi.org/10.1016/j.bdr.2015.03.003.
13. Mosavi, A.; Salimi, M.; Faizollahzadeh Ardabili, S.; Rabczuk, T.; Shamshirband, S.; Varkonyi-Koczy, A. State of the Art of Machine Learning Models in Energy Systems, a Systematic Review. *Energies (Basel)* **2019**, *12*(7), 1301. https://doi.org/10.3390/en12071301.
14. Xiao, L.; Wang, J.; Dong, Y.; Wu, J. Combined Forecasting Models for Wind Energy Forecasting: A Case Study in China. *Renewable and Sustainable Energy Reviews* **2015**, *44*, 271–288. https://doi.org/10.1016/j.rser.2014.12.012.
15. Dincer, I. Renewable Energy and Sustainable Development: A Crucial Review. *Renewable and Sustainable Energy Reviews* **2000**, *4*(2), 157–175. https://doi.org/10.1016/S1364-0321(99)00011-8.
16. Ahmad, T.; Zhang, D.; Huang, C.; Zhang, H.; Dai, N.; Song, Y.; Chen, H. Artificial Intelligence in Sustainable Energy Industry: Status Quo, Challenges and Opportunities. *Journal of Cleaner Production* **2021**, *289*, 125834. https://doi.org/10.1016/j.jclepro.2021.125834.

17. Rangel-Martinez, D.; Nigam, K. D. P.; Ricardez-Sandoval, L. A. Machine Learning on Sustainable Energy: A Review and Outlook on Renewable Energy Systems, Catalysis, Smart Grid and Energy Storage. *Chemical Engineering Research and Design* **2021**, *174*, 414–441. https://doi.org/10.1016/j.cherd.2021.08.013.
18. Bini, S. A. Artificial Intelligence, Machine Learning, Deep Learning, and Cognitive Computing: What Do These Terms Mean and How Will They Impact Health Care? *The Journal of Arthroplasty* **2018**, *33* (8), 2358–2361. https://doi.org/10.1016/j.arth.2018.02.067.
19. Elahi, M.; Afolaranmi, S. O.; Martinez Lastra, J. L.; Perez Garcia, J. A. A Comprehensive Literature Review of the Applications of AI Techniques through the Lifecycle of Industrial Equipment. *Discover Artificial Intelligence* **2023**, *3* (1), 43. https://doi.org/10.1007/s44163-023-00089-x.
20. Mocanu, E.; Nguyen, P. H.; Gibescu, M.; Kling, W. L. Deep Learning for Estimating Building Energy Consumption. *Sustainable Energy, Grids and Networks* **2016**, *6*, 91–99. https://doi.org/10.1016/j.segan.2016.02.005.
21. Serban, A. C.; Lytras, M. D. Artificial Intelligence for Smart Renewable Energy Sector in Europe—Smart Energy Infrastructures for Next Generation Smart Cities. *IEEE Access* **2020**, *8*, 77364–77377. https://doi.org/10.1109/ACCESS.2020.2990123.
22. Ozalp, H.; Cennamo, C.; Gawer, A. Disruption in Platform-Based Ecosystems. *Journal of Management Studies* **2018**, *55* (7), 1203–1241. https://doi.org/10.1111/joms.12351.
23. Sandström, C.; Berglund, H.; Magnusson, M. Symmetric Assumptions in the Theory of Disruptive Innovation: Theoretical and Managerial Implications. *Creativity and Innovation Management* **2014**, *23* (4), 472–483. https://doi.org/10.1111/caim.12092.
24. Inman, R. H.; Pedro, H. T. C.; Coimbra, C. F. M. Solar Forecasting Methods for Renewable Energy Integration. *Progress in Energy and Combustion Science* **2013**, *39* (6), 535–576. https://doi.org/10.1016/j.pecs.2013.06.002.
25. Sobri, S.; Koohi-Kamali, S.; Rahim, N. Abd. Solar Photovoltaic Generation Forecasting Methods: A Review. *Energy Conversion and Management* **2018**, *156*, 459–497. https://doi.org/10.1016/j.enconman.2017.11.019.
26. Raza, M. Q.; Nadarajah, M.; Ekanayake, C. On Recent Advances in PV Output Power Forecast. *Solar Energy* **2016**, *136*, 125–144. https://doi.org/10.1016/j.solener.2016.06.073.
27. McGovern, A.; Elmore, K. L.; Gagne, D. J.; Haupt, S. E.; Karstens, C. D.; Lagerquist, R.; Smith, T.; Williams, J. K. Using Artificial Intelligence to Improve Real-Time Decision-Making for High-Impact Weather. *Bulletin of the American Meteorological Society* **2017**, *98* (10), 2073–2090. https://doi.org/10.1175/BAMS-D-16-0123.1.
28. Jha, S. Kr.; Bilalovic, J.; Jha, A.; Patel, N.; Zhang, H. Renewable Energy: Present Research and Future Scope of Artificial Intelligence. *Renewable and Sustainable Energy Reviews* **2017**, *77*, 297–317. https://doi.org/10.1016/j.rser.2017.04.018.
29. Seagraves, P. Real Estate Insights: Is the AI Revolution a Real Estate Boon or Bane? *Journal of Property Investment & Finance* **2023**. https://doi.org/10.1108/JPIF-05-2023-0045.

30. Kabaivanov, S.; Markovska, V. Artificial Intelligence in Real Estate Market Analysis; 2021; p 030001. https://doi.org/10.1063/5.0041806.
31. GINA CRISTINA, D.; ERIKA, M.; JOSEF, J. Investigating the Long and Short-Run Salary- Employment Relationship in Romania: A Sectorial Approach Using the ARDL Model. *Econ Comput Econ Cybern Stud Res* **2019**, *53* (1/2019), 5–20. https://doi.org/10.24818/18423264/53.1.19.01.
32. Ssekulima, E. B.; Anwar, M. B.; Al Hinai, A.; El Moursi, M. S. Wind Speed and Solar Irradiance Forecasting Techniques for Enhanced Renewable Energy Integration with the Grid: A Review. *IET Renewable Power Generation* **2016**, *10* (7), 885–989. https://doi.org/10.1049/iet-rpg.2015.0477.
33. White, W.; Lunnan, A.; Nybakk, E.; Kulisic, B. The Role of Governments in Renewable Energy: The Importance of Policy Consistency. *Biomass Bioenergy* **2013**, *57*, 97–105. https://doi.org/10.1016/j.biombioe.2012.12.035.
34. Zeng, Y.; Dong, P.; Shi, Y.; Li, Y. On the Disruptive Innovation Strategy of Renewable Energy Technology Diffusion: An Agent-Based Model. *Energies (Basel)* **2018**, *11* (11), 3217. https://doi.org/10.3390/en11113217.
35. Lytras, M.; Aljohani, N. R.; Hussain, A.; Luo, J.; Zhang, J. X. Cognitive Computing Track Chairs' Welcome & Organization. In *Companion of the The Web Conference 2018 on The Web Conference 2018 - WWW '18*; ACM Press: New York, USA, **2018**; pp 247–250. https://doi.org/10.1145/3184558.3192295.
36. Lytras, M. D.; Raghavan, V.; Damiani, E. Big Data and Data Analytics Research. *International Journal on Semantic Web and Information* **2017**, *13*(1), 1–10. https://doi.org/10.4018/IJSWIS.2017010101.
37. Jain, N.; Shashi Bhushan, B. L.; Natarajan, M.; Mehta, R.; Saini, D. K.; Chatterjee, K. Advanced 3D *In Vitro* Lung Fibrosis Models: Contemporary Status, Clinical Uptake, and Prospective Outlooks. *ACS Biomaterials Science & Engineering Journal* **2024**, *10* (3), 1235–1261. https://doi.org/10.1021/acsbiomaterials.3c01499.

Index

Note: **Bold** page numbers refer to tables and *italic* page numbers refer to figures.

Printed in the United States
by Baker & Taylor Publisher Services